SOTHIS PRESS

Sothis Press aims to completely reimagine the public domain book with an emphasis on universal access to captivatingly designed print editions. We believe that technology will enable every single book ever published to be perpetually accessible in beautiful print form. Sothis is beauty in ubiquity.

www.sothis.press
1 Waterhouse Square, London, ECIN 2ST

The Analysis of Paints

And Painting Materials

Henry A. Gardner

SOTHIS PRESS

ISBN 978-1-0368-0991-1

This edition is published by
Sothis Press © 2023. The company
is registered in England and Wales.
The Company Number is 14900389.

THE ANALYSIS OF PAINTS

AND

PAINTING MATERIALS

Published by the
McGraw-Hill Book Company
New York

Successors to the Book Departments of the
McGraw Publishing Company Hill Publishing Company

Publishers of Books for
Electrical World
The Engineering and Mining Journal
Engineering Record
Power and The Engineer
Electric Railway Journal
American Machinist
Metallurgical and Chemical Engineering

THE ANALYSIS OF PAINTS

AND

PAINTING MATERIALS

BY

HENRY A. GARDNER

DIRECTOR, SCIENTIFIC SECTION, EDUCATIONAL BUREAU, PAINT MANUFACTURERS' ASSOCIATION OF THE UNITED STATES

AND

JOHN A. SCHAEFFER

INSTRUCTOR IN CHEMICAL PRACTICE, CARNEGIE TECHNICAL SCHOOLS, PITTSBURG, PA.

McGRAW-HILL BOOK COMPANY

239 WEST 39TH STREET, NEW YORK

6 BOUVERIE STREET, LONDON, E. C.

1911

Printed by
The Maple Press
York, Pa.

To

Dr. Edgar F. Smith

Provost-elect, University of Pennsylvania, whose teachings have been a source of inspiration in our work, this book is dedicated.

PREFACE.

The authors are presenting in this book a series of selected methods for the analysis of materials used in the manufacture of paints.

Acknowledgment is made to Walker, McIlhiney, and others for several methods of importance, which have been included and correlated with new and valuable methods worked out by the authors.

It is assumed that the reader is well versed in the ordinary quantitative methods used in analytical chemistry, and no attempt, therefore, has been made to explain such methods in detail.

It is the hope of the authors that this book will prove of value to all those engaged in the manufacture or use of painting materials.

December, 1910.

CONTENTS.

THE ANALYSIS OF PAINTS

AND

PAINTING MATERIALS

CHAPTER I.

THE ANALYSIS OF DRY PIGMENTS.

ZINC OXIDE.

Zinc Oxide.—Zinc oxide contains approximately 80 per cent. of metallic zinc, the balance being combined oxygen. This pigment is completely soluble in acetic acid. Some varieties, however, contain a small percentage of impurities, such as lead sulphate, zinc sulphate, sulphur dioxide, silica, iron, and traces of metallic zinc. This pigment may be analyzed either gravimetrically or volumetrically.

Weigh 1 gram into a beaker, dissolve in acetic acid, filter off any insoluble residue and determine the percentage of zinc in the filtrate by one of the following methods:

1. *Gravimetric Method.*—Heat the acetic acid filtrate to boiling. Completely precipitate the zinc with hydrogen sulphide, and boil for ten minutes, having the solution at the end of the operation smelling strongly of hydrogen sulphide. Filter, wash, and dissolve the precipitate in hydrochloric acid. Cool, add sodium carbonate solution drop by drop until the solution becomes turbid. Heat to boiling, add 2 drops of phenolphthalein solution and continue the addition of sodium carbonate solution until just alkaline, when the zinc will be completely precipitated as carbonate. Filter in a Gooch crucible, while still hot wash, ignite and weigh as zinc oxide.

2. *Volumetric Method.*—Treat the acetic acid filtrate with ammonium hydroxide until alkaline. Then add hydrochloric acid until faintly acid. Three c.c. of concentrated hydrochloric acid in excess are then added, the solution diluted to 250 c.c., and

titrated with standard potassium ferrocyanide as in the standardization of the solution. This method is not applicable in the presence of iron or manganese.

Standardization of the Ferrocyanide Solution.—Ten grams of pure metallic zinc are carefully weighed off and dissolved in hydrochloric acid. The solution is made up to 1 liter and a volume equivalent to .2 gram is measured out. The remaining solution may be kept for restandardizing the ferrocyanide solution which, on standing, appears to change from time to time. In place of using the standard zinc solution, .2 gram of pure metallic zinc may be used for each standardization.

The ferrocyanide solution is made by dissolving 22 grams of crystallized potassium ferrocyanide in a liter of water. One c.c. of this solution will be equal to about .005 gram of metallic zinc.

The indicator is prepared by dissolving uranium acetate or uranium nitrate in water until a faint yellow color is produced. A 5 per cent. solution will usually give a good end reaction. A number of drops of this solution are placed on a spot plate, and the end point determined by introducing a few drops of the solution which is being titrated.

The acid solution containing .2 gram of zinc is made faintly alkaline with ammonium hydroxide, using litmus paper to determine the end point. Reacidify faintly with hydrochloric acid and add 3 c.c. of concentrated hydrochloric acid in excess. Dilute to about 250 c.c. and heat to about 80° C.

The hot zinc solution is divided into two equal portions. One portion is titrated by slowly running in a few c.c. of ferrocyanide solution at a time, with vigorous stirring, until a few drops of the solution give a brownish tinge to the uranium acetate indicator on the spot plate. The remainder of the zinc solution, with the exception of a few c.c., is now added to the titrated solution and the end point again determined. The titrated solution is now poured back into the original beaker containing the few remaining c.c. of untitrated solution. The titration is finished by adding two drops of the ferrocyanide solution at a time, testing for the end point after each addition. As the end point develops slowly, it is well to examine each spot, after standing for a brief time. The first one developing a brown tinge is taken as the end point.

A blue color will appear in the hot zinc solution upon the addition of the ferrocyanide solution. This color gradually becomes lighter, and when the end point is reached changes to a white. By repeated titrations, this end point can be easily noted and serves to shorten the time required for the outside testing with uranium acetate. It, however, should not be taken as final, as the uranium acetate gives a more definite end point. The blue color will not be present in an excess of ferrocyanide solution.

It is necessary to make a correction for the amount of ferrocyanide solution required to develop a brown color in the uranium acetate indicator when zinc is absent. This correction is deducted from the total amount of ferrocyanide solution used and will usually run between .1 and .2 c.c.

BASIC CARBONATE—WHITE LEAD.

Basic carbonate white lead ($2PbCO_3Pb(OH)_2$) contains approximately 80 per cent. metallic lead and 20 per cent. carbonic acid and combined water, with traces sometimes of silver, antimony, lead, and other metals. The analysis of white lead can best be carried out by Walker's* method.

(a) Total Lead.

"Weigh 1 gram of the sample, moisten with water, dissolve in acetic acid, filter, and wash, ignite, and weigh the insoluble impurities. To the filtrate from the insoluble matter add 25 c.c. of sulphuric acid (1:1), evaporate and heat until the acetic acid is driven off; cool, dilute to 200 c.c. with water, add 20 c.c. of ethyl alcohol, allow to stand for two hours, filter on a Gooch crucible, wash with 1 per cent. sulphuric acid, ignite, and weigh as lead sulphate. Calculate to total lead ($PbSO_4 \times 0.68292 = Pb$), or calculate to basic carbonate of lead (white lead) by multiplying the weight of lead sulphate by 0.85258.

"The filtrate from the lead sulphate may be used to test for other metals, though white lead is only ra ely adulterated with soluble substances; test, however, for zinc, which may be present as zinc oxide.

* P. H. Walker, Bureau of Chemistry Bulletin No. 109, revised, U. S. Dept. of Agriculture, pp. 21 and 22.

"Instead of determining the total lead as sulphate it may be determined as lead chromate by precipitating the hot acetic acid solution with potassium dichromate, filtering on a Gooch crucible, igniting at a low temperature, and weighing as lead chromate.

(b) Complete Analysis.

"When it is necessary to determine the exact composition of a pure white lead, heat 1 gram of the pigment in a porcelain boat in a current of dry, carbon-dioxide-free air, catching the water in sulphuric acid and calcium chloride and the carbon dioxide in soda lime or potassium hydroxide (1.27 specific gravity). By weighing the residue of lead monoxide in the boat all the factors for determining the total composition are obtained. Figure the carbon dioxide to lead carbonate ($PbCO_3$), calculate the lead monoxide corresponding to the lead carbonate ($PbCO_3$) and subtract from the total lead monoxide, calculate the remaining lead monoxide to lead hydroxide ($Pb(OH)_2$), calculate the water corresponding to lead hydroxide and subtract from the total water, the remainder being figured as moisture.

"This method assumes the absence of acetic acid. Thompson* states that acetic acid varies from 0.05 per cent. in Dutch process white lead to 0.7 per cent. in some precipitated white leads. It is then more accurate to determine the carbon dioxide by evolution; this is especially the case when working with a lead extracted from an oil paste, as the lead soap and unextracted oil will cause a considerable error by the ignition method. In determining carbon dioxide by the evolution method, liberate the carbon dioxide with dilute nitric acid, have a reflux condenser next to the evolution flask and dry the carbon dioxide with calcium chloride before absorbing it in the potassium hydroxide bulbs.

(c) Acetic Acid.

"It is sometimes necessary to determine acetic acid. The Navy Department specifications demand that white lead shall not contain 'acetate in excess of fifteen one-hundredths of 1 per cent. of glacial acetic acid.' Thompson's method* is as follows:

*J. Soc. Chem. Ind., 1905, 24: 487.

"'Eighteen grams of the dry white lead are placed in a 500 c.c. flask, this flask being arranged for connection with a steam supply and also with an ordinary Liebig condenser. To this white lead is added 40 c.c. of syrupy phosphoric acid, 18 grams of zinc dust, and about 50 c.c. of water. The flask containing the material is heated directly and distilled down to a small bulk. Then the steam is passed into the flask until it becomes about half-full of condensed water, when the steam is shut off and the original flask heated directly and distilled down to the same small bulk—this operation being conducted twice. The distillate is then transferred to a special flask and 1 c.c. of syrupy phosphoric acid added to ensure a slightly acid condition. The flask is then heated and distilled down to a small bulk—say 20 c.c. Steam is then passed through the flask until it contains about 200 c.c. of condensed water, when the steam is shut off and the flask heated directly. These operations of direct distillation and steam distillation are conducted until 10 c.c. of the distillate require but a drop of tenth-normal alkali to produce a change in the presence of phenolphthalein. Then the bulk of the distillate is titrated with tenth-normal sodium hydroxide, and the acetic acid calculated. It will be found very convenient in this titration, which amounts in some cases to 600–700 c.c. to titrate the distillate when it reaches 200 c.c., and so continue titrating every 200 c.c. as it distils over.'

"If the white lead contains appreciable amounts of chlorine it is well to add some silver phosphate to the second distillation flask and not carry the distillation from this flask too far at any time.

"The method used by the chemists of the Navy Department is as follows: Weigh 25 grams of white lead in an Erlenmeyer flask, add 75 c.c. of 25 per cent. phosphoric acid, distil with steam to a 500 c.c. distillate, add to the distillate some milk of barium carbonate, bring to a boil, filter, keeping the solution at the boiling point (it is not necessary to wash), add an excess of sulphuric acid to the filtrate and determine the barium sulphate in the usual manner; subtract 53 milligrams from the weight of the barium sulphate and calculate the remainder as acetic acid ($BaSO_4 \times 0.515 = CH_3COOH$). The object of this rather indirect method is to avoid any error that might arise from fatty acids being carried over by the steam distillation. For white lead that has not been ground in oil, Thompson's method is to be preferred."

Volumetric Method.—The following volumetric method for the determination of lead has been found by the authors to give excellent results when the precautions given are carefully observed.

A .5 gram sample is dissolved in 10 c.c. of concentrated hydrochloric acid and boiled until solution is effected. Cool, dilute to 40 c.c., neutralize with ammonium hydroxide. Add acetic acid until distinctly acid. Dilute to 200 c.c. with hot water. Boil and titrate with ammonium molybdate as given in the standardization of ammonium molybdate.

Standardization of Ammonium Molybdate.—Four and one-fourth grams of ammonium molybdate are dissolved to the liter, so that each c.c. is equivalent to 1 per cent. of lead when .5 gram sample is taken. Standardize with .2 gram pure lead foil. Dissolve the lead in nitric acid, evaporate nearly to dryness, add 30 c.c. of water, then 5 c.c. sulphuric acid specific gravity 1.84, cool, and filter. Drop the filter containing the precipitated lead sulphate into a flask, add 10 c.c. hydrochloric acid, sp. gr. 1.19, boil to complete disintegration, add 15 c.c. of hydrochloric acid, 25 c.c. of water, and ammonium hydroxide until alkaline. Make acid with acetic acid and dilute to 200 c.c. with hot water and boil. Titrate, using an outside indicator of one part of tannic acid in 300 parts of water.

The following precautions must be observed in carrying out this method. Calcium forms a molybdate more or less insoluble, and when calcium is present, results are apt to be high. However, when less than 2 per cent. of calcium is present and a high percentage of lead, there appears to be no interference from the calcium. This method is only good for samples containing more than 10 per cent. of lead. Should a lower percentage of lead be present, it must be precipitated as the sulphate then redissolved and titrated as in the method of standardization.

Carbonic Acid.—The carbonic-acid content of white lead may be determined by using the Scheibler apparatus, as follows:

One gram of the dry pigment is placed in the small tube (B) contained in the flask (F). Dilute hydrochloric acid is placed in the flask (F) on the outside of the small tube. Water is brought above the zero mark in the tube (M), by forcing water from the flask (E) by means of the bulb (A). On opening the

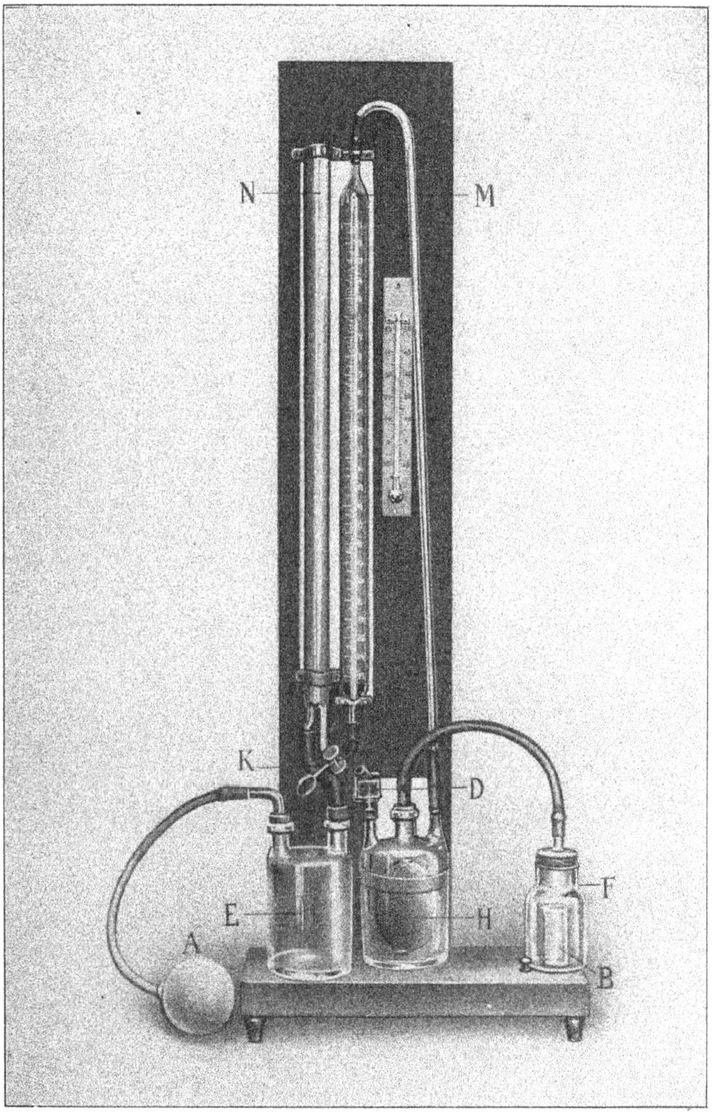

FIG. 1.—Scheibler Apparatus.

stopcock (D), water is allowed to reach a level on the zero mark in the two tubes (M) and (N). The flask (F) is then inverted and the acid is allowed to act on the sample. The gas evolved passes into the rubber balloon (H) which causes the water in the tube (M) to lower and that in (N) to rise. The pinchcock (K) is opened and the water is allowed to flow out of the tube (N) at such a rate so that the liquid in the two tubes will be on the same level, as nearly as possible. When the gas ceases to come off, the pinchcock (K) is closed and the apparatus allowed to stand for several minutes, after which the flask (F) is shaken several times so as to complete the action. After action has ceased, which should take place in from a half to three-quarters of an hour, the water in the two tubes is brought to the same level and the amount of gas which has been evolved is noted. It is essential that temperature and barometric readings be made at the same time, so that corrections may be made for errors arising from these factors. The calculation is made directly by referring to the volume of carbon dioxide evolved at the given temperature and pressure in the tables which are given for this conversion, making correction also for the absorption of carbon dioxide in dilute hydrochroric acid, the tables taking directly into account the errors arising from temperature and pressure. In some cases where calcium carbonate is present in a pigment made from dolomitic limestones, it is necessary that the contents of the flask be boiled, although this need be done only in rare cases.

The following Dietrich tables are so arranged that the conversions can easily be made.

TABLE I.

Absorption of Carbonic Acid Gas, in Cubic Centimeters of CO_2, using 10 c.c. of Dilute Hydrochloric Acid (1.10 Sp. Gr.).

C. C. evolved	1	2	3	4	5	6	7	8	9	10
C. C. sbed	1.85	2.00	2.16	2.31	2.47	2.62	2.78	2.93	3.09	3.24
C. C. dul	11	12	13	14	15	16	17	18	19	20
C. C. absorbed	3.40	3.55	3.71	3.86	4.02	4.17	4.33	4.48	4.64	4.79
C. C. bed	21	22	23	24	25	26	27	28	29	30
C. C. absorbed	4.95	4.96	4.97	4.98	5.00	5.03	5.04	5.06	5.07	5.09
C. C. evolved	31	32	33	34	35	36	37	38	39	40
C. C. absorbed	5.10	5.11	5.13	5.14	5.16	5.17	5.18	5.20	5.21	5.23
C. C. bed	41	42	43	44	45	46	47	48	49	50
C. C. absorbed	5.24	5.25	5.26	5.27	5.29	5.30	5.31	5.32	5.34	5.35
C. C. evolved	51	52	53	54	55	56	57	58	59	60
C. C. absorbed	5.36	5.37	5.38	5.40	5.41	5.43	5.44	5.45	5.47	5.48
C. C. evolved	61	62	63	64	65	66	67	68	69	70
C. C. absorbed	5.50	5.51	5.52	5.54	5.55	5.57	5.58	5.59	5.61	5.62
C. C. evolved	71	72	73	74	75	76	77	78	79	80
C. C. absorbed	5.64	5.65	5.66	5.68	5.69	5.71	5.72	5.73	5.75	5.76
C. C. dued	81	82	83	84	85	86	87	88	89	90
C. C. absorbed	5.78	5.79	5.80	5.82	5.83	5.85	5.86	5.87	5.89	5.90
C. C. evolved	91	92	93	94	95	96	97	98	99	100
C. C. absorbed	5.92	5.93	5.94	5.96	5.97	5.99	6.00	6.01	6.03	6.04

TABLE II.

Weight in Decimal Parts of a Gram of One Cubic Centimeter of Carbonic Acid Gas (CO_2).—Dietrich Tables.

C°.	720 mm. 28.35 in.	722 mm. 28.42 in.	724 mm. 28.50 in.	726 mm. 28.58 in.	728 mm. 28.66 in.	730 mm. 28.74 in.	732 mm. 28.82 in.
10°	.001774	.001779	.001784	.001789	.001794	.001799	.001804
1	.001767	.001772	.001777	.001782	.001786	.001791	.001796
12	.001759	.001764	.001769	.001774	.001779	.001783	.001788
13	.001751	.001756	.001761	.001766	.001771	.001776	.001780
14	.001743	.001748	.001753	.001758	.001763	.001768	.001773
15	.001735	.001740	.001745	.001750	.001755	.001759	.001764
16	.001727	.001732	.001737	.001742	.001746	.001751	.001756
17	.001719	.001724	.001729	.001733	.001738	.001743	.001748
18	.001711	.001715	.001720	.001725	.001730	.001735	.001740
19	.001702	.001707	.001712	.001717	.001722	.001727	.001731
20	.001694	.001699	.001704	.001708	.001713	.001718	.001723
21	.001686	.001690	.001695	.001700	.001705	.001710	.001714
22	.001677	.001682	.001687	.001691	.001696	.001701	.001706
23	.001669	.001673	.001678	.001683	.001688	.001692	.001697
24	.001660	.001665	.001669	.001674	.001679	.001684	.001688
25	.001651	.001656	.001661	.001665	.001670	.001675	.001679
26	.001642	.001647	.001652	.001656	.001661	.001666	.001670
27	.001633	.001638	.001643	.001647	.001652	.001657	.001661
28	.001624	.001629	.001634	.001638	.001643	.001648	.001652
29	.001615	.001620	.001625	.001629	.001634	.001639	.001643
30	.001606	.001611	.001616	.001620	.001625	.001630	.001634

TABLE II.—Continued.

C°.	734 mm. 28.90 in.	736 mm. 28.98 in.	738 mm. 29.05 in.	740 mm. 29.13 in.	742 mm. 29.21 in.	744 mm. 29.29 in.	746 mm. 29.37 in.
10°	.001809	.001814	.001819	.001824	.001829	.001834	.001839
11	.001801	.001806	.001811	.001816	.001821	.001826	.001831
12	.001793	.001798	.001803	.001808	.001813	.001818	.001823
13	.001785	.001790	.001795	.001800	.001805	.001810	.001815
14	.001777	.001782	.001787	.001792	.001797	.001802	.001807
15	.001769	.001774	.001779	.001784	.001789	.001794	.001799
16	.001761	.001766	.001771	.001776	.001781	.001786	.001791
17	.001753	.001758	.001763	.001767	.001772	.001777	.001782
18	.001745	.001749	.001754	.001759	.001764	.001769	.001774
19	.001736	.001741	.001746	.001751	.001756	.001760	.001765
20	.001728	.001733	.001737	.001742	.001747	.001752	.001757
21	.001719	.001724	.001729	.001734	.001739	.001743	.001748
22	.001711	.001715	.001720	.001725	.001730	.001735	.001739
23	.001702	.001707	.001712	.001716	.001721	.001726	.001731
24	.001693	.001698	.001703	.001707	.001712	.001717	.001722
25	.001684	.001689	.001694	.001698	.001703	.001708	.001713
26	.001675	.001680	.001685	.001689	.001694	.001699	.001704
27	.001666	.001671	.001676	.001680	.001685	.001690	.001695
28	.001657	.001662	.001667	.001671	.001676	.001681	.001686
29	.001648	.001653	.001658	.001662	.001667	.001672	.001677
30	.001639	.001644	.001649	.001653	.001658	.001663	.001668

TABLE III.

Weight in Decimal Parts of a Gram of One Cubic Centimeter of Carbonic Acid Gas (CO_2).—Dietrich Tables (Contd.).

C°.	748 mm. 29.45 in.	750 mm. 29.53 in.	752 mm. 29.61 in.	754 mm. 29.68 in.	756 mm. 29.76 in.	758 mm. 29.84 in.	760 mm. 29.92 in.
10°	.001844	.001849	.001854	.001859	.001864	.001869	.001874
11	.001836	.001841	.001846	.001851	.001856	.001861	.001866
12	.001828	.001833	.001838	.001843	.001848	.001853	.001858
13	.001820	.001825	.001830	.001835	.001840	.001845	.001850
14	.001812	.001817	.001822	.001827	.001832	.001837	.001842
15	.001804	.001809	.001813	.001818	.001823	.001828	.001833
16	.001795	.001800	.001805	.001810	.001815	.001820	.001825
17	.001787	.001792	.001797	.001802	.001806	.001811	.001816
18	.001779	.001783	.001788	.001793	.001798	.001803	.001808
19	.001770	.001775	.001780	.001785	.001789	.001794	.001799
20	.001761	.001766	.001771	.001776	.001781	.001786	.001791
21	.001753	.001758	.001763	.001767	.001772	.001777	.001782
22	.001744	.001749	.001754	.001759	.001763	.001768	.001773
23	.001735	.001740	.001745	.001750	.001754	.001759	.001764
24	.001726	.001731	.001736	.001741	.001745	.001750	.001755
25	.001717	.001722	.001727	.001732	.001736	.001741	.001746
26	.001708	.001713	.001718	.001723	.001727	.001732	.001737
27	.001699	.001704	.001709	.001714	.001718	.001723	.001728
28	.001690	.001695	.001700	.001705	.001709	.001714	.001719
29	.001681	.001686	.001691	.001696	.001700	.001705	.001710
30	.001672	.001677	.001682	.001687	.001691	.001696	.001701

TABLE III.—Continued.

C°.	762 mm. 30.00 in.	764 mm. 30.08 in.	766 mm. 30.16 in.	768 mm. 30.24 in.	770 mm. 30.32 in.	772 mm. 30.40 in.	774 mm. 30.48 in.
10°	.001879	.001884	.001889	.001894	.001899	.001904	.001909
11	.001871	.001876	.001881	.001886	.001891	.001896	.001901
12	.001863	.001868	.001873	.001878	.001883	.001888	.001893
13	.001855	.001860	.001865	.001870	.001874	.001879	.001884
14	.001847	.001851	.001856	.001861	.001866	.001871	.001876
15	.001838	.001843	.001848	.001853	.001858	.001863	.001868
16	.001830	.001835	.001839	.001844	.001849	.001854	.001859
17	.001821	.001826	.001831	.001836	.001841	.001846	.001851
18	.001813	.001817	.001823	.001827	.001832	.001837	.001842
19	.001804	.001809	.001814	.001818	.001823	.001828	.001833
20	.001795	.001800	.001805	.001810	.001815	.001820	.001825
21	.001787	.001791	.001796	.001801	.001806	.001811	.001816
22	.001778	.001783	.001787	.001792	.001797	.001802	.001807
23	.001769	.001774	.001778	.001783	.001788	.001793	.001798
24	.001760	.001765	.001769	.001774	.001779	.001784	.001789
25	.001751	.001755	.001760	.001765	.001770	.001775	.001780
26	.001742	.001746	.001751	.001756	.001761	.001766	.001771
27	.001733	.001737	.001742	.001747	.001752	.001757	.001762
28	.001724	.001728	.001733	.001738	.001743	.001748	.001753
29	.001715	.001719	.001724	.001729	.001734	.001739	.001744
30	.001706	.001710	.001715	.001720	.001725	.001730	.001735

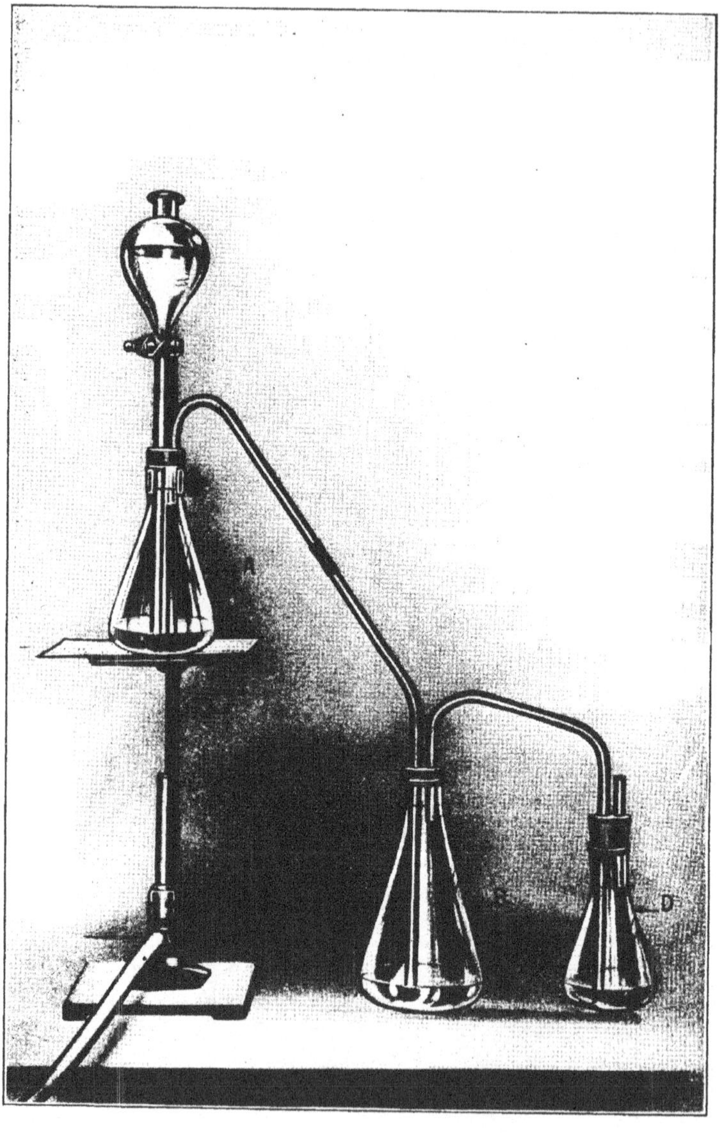

FIG. 2.—Carbon Dioxide Apparatus.

The authors and de Horvath have developed the following method as a simple and more efficacious means of determining the carbonic-acid content of paint pigments.

The method can be used in such cases where the substances to be analyzed evolve gases other than carbon dioxide; that is, hydrogen sulphide, sulphur dioxide, or organic matter. The apparatus used is shown in the accompanying diagram. A weighed sample of the substance is introduced into the Erlenmeyer flask (A). Into flask (B) is placed a 10 per cent. solution of barium chloride, more than sufficient to hold the carbon dioxide evolved, and 20 c.c. of concentrated ammonium hydroxide free from carbon dioxide. If sulphides are present, it is sometimes advisable to pass the liberated gas first through a few c.c. of strong potassium permanganate. The flask (B) is warmed until completely filled with ammonia fumes. Flask (D) is a safety bottle containing the same solution as flask (B). Only in rare cases will any trace of carbon dioxide be noticed in the safety flask. After flask (B) is completely filled with ammonia vapor, make all connections and allow the hydrochloric acid to drop slowly from the separatory funnel into the decomposition flask (A). When effervescence has ceased, heat the contents of the flask until filled with steam. The delivery tubes and sides of the precipitating flask are then washed with boiling water, the flask is filled to the neck, stoppered, and the precipitated barium carbonate allowed to settle. Wash thoroughly by decantation, each time stoppering the flask to prevent any error from the carbon dioxide present in the air, and determine either gravimetrically, by conversion into barium sulphate, or volumetrically, by dissolving in standard hydrochloric acid and titrating the excess of acid used with standard potassium hydroxide. Calculate the barium found to carbonate and the amount of carbon dioxide from the found carbonate. The entire operation may be hastened by conducting a brisk current of air free from carbon dioxide through the entire apparatus.

A few typical analyses of white lead, for impurities, are given.

Table IV.—Typical Analyses of White Lead for Impurities.

	1	2	3	4	5	6	7	8
Insoluble	.0252%	.0130%	.0241%	.0417%	.0089%	.0307%	.0123%	.0594%
Antimony	.0019	.0085	.0035	.0028	.0031	.0074	.0054	.0080
Arsenic	.0001	.0072	.0010	.0024	.0022	.0027	.0018	.0008
Bismuth	.0197	.0165	.0122	.0209	.0329	.0226	.0395	.0133
Silver	none	none	none	none	none	none	none	none
Cadmium	none	none	none	none	none	none	none	none
Copper	.0005	trace	.0014	.0012	.0013	.0008	. 06	.0111
Nickel	none	trace	trace	none	none	none	trace	none
Zinc	.0008	.0005	. 06	.0008	. 09	.0003	.0005	.0003
Manganese	none	none	none	none	none	none	none	none
Calcium	.0300	.0057	.0142	.0071	.0143	.0143	.0100	.0172
Iron oxide	.0038	.0045	.0042	.0035	.0043	.0031	.0035	.0056
Aluminum oxide	. 09	.0005	.0004	.00 4	.0001	. 04	.0019	.0032
Sulphur trioxide	.0338	.0021	.0105	.0049	.0442	.0049	.0099	.0196
Chlorine	.0299	.0497	.0165	.0403	.0149	.0420	.0425	.01 8
Acetic acid	.4539	.3157	.3060	.2558	.2558	.5462	.5853	.2681

Metallic lead which comes to the corroder may contain very minute traces of metals, such as silver, copper, bismuth, cadmium, antimony, arsenic, iron, nickel, copper, zinc, and manganese. Should an analysis of metallic lead be necessary, it can be best carried out by the methods outlined by Fresenius' Quantitative Chemical Analysis, Vol. II, pp. 584 to 593. The method is here omitted owing to the large number of references which must be necessarily followed in making an examination for the above impurities.

Basic Sulphate—White Lead (Sublimed White Lead).—Basic sulphate white lead shows approximately 70 to 75 per cent. of lead sulphate, 20 per cent. lead oxide, and 5 per cent. zinc oxide. The Hughes* method for the analysis of this pigment has, by long experience, been found to give the best results.

Total Sulphates.—Mix in a beaker 1/2 gram sample with 3 grams sodium carbonate. Add 30 c.c. of water and boil gently for ten minutes. Allow to stand for four hours. Dilute with hot water, filter, and wash until filtrate is about 200 c.c. in volume. Reject the residue. By this reaction all the lead sulphate is changed to carbonate and the filtrate contain sodium sulphate.

Acidulate the filtrate with hydrochloric acid. Boil, and add a slight excess of barium chloride solution. When the precipitate has well settled, filter on an ashless filter, wash, ignite, and weigh as $BaSO_4$.

Lead (*Method* 1).—Dissolve 1 gram of sample in 100 c.c. of a solution made as follows:

Eighty per cent. acetic acid,	125 c.c.
Concentrated ammonia,	95 c.c.
Water,	100 c.c.

Add this solution hot and dilute with about 50 c.c. of water. Boil until dissolved.

Dilute to 200 c.c. and titrate with standard ammonium molybdate solution, spotting out on a freshly prepared solution of tannic acid. The details of this method are given under the analysis of basic carbonate white lead.

Ammonium molybdate is a slightly variable salt, but a solution containing 8.67 grams per liter usually gives a standard solution:

* L. S. Hughes, chief chemist, Picher Lead Co.

1 c.c. equals .01 gram Pb.

Standardize against pure PbO or pure $PbSO_4$.

Lead (*Method* 2).—Treat sample as above until dissolved. If solution is not quite clear, filter. Add to filtrate an excess of potassium bichromate solution. Boil and stand in a warm place until well settled. Filter on a Gooch crucible, ignite below a red heat and weigh as $PbCrO_4$.

Factor for lead equals .64.

Zinc.—Boil 1 gram of sample in a beaker with the following solution:

Water,	30 c.c.
Ammonium chloride,	4 gms.
Concentrated hydrochloric acid,	6 c.c.

If sample is not quite dissolved, the result is not affected, as the residue is lead sulphate or precipitated lead chloride.

Dilute to 200 c.c. with hot water and titrate with a standard solution of potassium ferrocyanide, spotting out on a 5 per cent. solution of uranium nitrate.

Forty-three grams per liter of potassium ferrocyanide usually gives a solution approximately—1 c.c. equals .01 gram Zn. Standardize against freshly ignited ZnO or pure metallic zinc.

Sulphur Dioxide.—Digest 2 grams with frequent stirring in 5 per cent. sulphuric acid for ten minutes in the cold.

Add starch indicator and titrate with N/100 iodine solution.

A more accurate method is to add an excess of the standard iodine solution to the sample before addition of the acid and then to titrate the excess of iodine with N/100 sodium thiosulphate solution.

Calculations.—Calculations can be simplified by bearing in mind the following relations:

Weight of $BaSO_4 \times 1.3 =$ weight $PbSO_4$.

Deduct from the total lead the lead represented by the lead sulphate and calculate the residual lead to PbO.

Calculate the Zn found to ZnO.

Report the sulphur dioxide found directly, as it does not exist as a sulphite, but is instead an apparently occluded gas.

Zinc Lead and Leaded Zinc.

Zinc lead is approximately 50 per cent. zinc oxide and 50 per cent. lead sulphate. Leaded zinc varies considerably in its composition, though it contains on an average 25 per cent. lead sulphate and 75 per cent. zinc oxide. Both are apt to contain traces of zinc sulphate, sulphur dioxide, arsenic, and antimony. The analysis of such compounds is carried out in the following way:

Moisture.—Dry 2 grams of the sample at 105° C. for two hours. The loss will be moisture.

Lead and Zinc.—Treat one gram of the sample in a beaker with 10 c.c. of water, 10 c.c. of concentrated hydrochloric acid, and 5 grams of ammonium chloride. Heat on the steam bath for a few minutes, dilute to about 300 c.c., boil, filter, wash, and weigh any insoluble residue. In a pure leaded zinc or zinc lead compound no residue should remain. Examine the residue, should any be present. The lead is completely precipitated in the filtrate with hydrogen sulphide in the cold. Allow to settle, filter, wash, dissolve in nitric acid, evaporate in the presence of sulphuric acid, and determine the lead either volumetrically or gravimetrically, as stated in the analysis of basic carbonate white lead. The filtrate from the lead sulphide precipitate is made alkaline with ammonium hydroxide, any precipitate forming, due to the presence of iron, being filtered off, redissolved in hydrochloric acid, oxidized with a little nitric acid, reprecipitated with ammonia, washed and weighed. The ammoniacal filtrate is made slightly acid with acetic acid, heated to boiling, and the zinc is precipitated as zinc sulphide, filtered, washed, and determined either gravimetrically or volumetrically, as stated under the analysis of zinc oxide. The operation may be hastened by titrating the zinc directly after the removal of the lead, as stated in the volumetric determination of the zinc. The filtrate, after precipitating the zinc as sulphide, is examined for calcium and magnesium in the usual way. Should calcium or magnesium be present, it is, of course, understood that the zinc must first be removed as sulphide before titration in order to determine the calc'um and magnesium present.

Soluble Sulphate.—One gram of the sample is boiled with 100 c.c. of a mixture containing one part alcohol and three parts of water. Filter and wash with a similar mixture of alcohol and water, and determine the sulphuric acid in the filtrate by precipitation with barium chloride as barium sulphate. The sulphate thus found is calculated to zinc sulphate. Any zinc remaining after the calculation to zinc sulphate is calculated to zinc oxide and reported as such.

Total Sulphate.—One gram of the sample is treated with 10 c.c. of water, 10 c.c. of strong hydrochloric acid, and 5 grams of ammonium chloride, in a large beaker. Heat in a steam bath for a few minutes, dilute with hot water to about 300 c.c., boil, and filter off any insoluble residue. Heat the filtrate to boiling and determine the sulphate present, by precipitation with barium chloride as barium sulphate, in the usual way. Calculate the total sulphate present, deduct from it the soluble sulphate, and calculate the remainder to lead sulphate. Any lead remaining after the calculation to lead sulphate is calculated to lead monoxide.

The authors find the iodine test for the determination of SO_2 in zinc lead and leaded zinc, as shown under the analysis of sublimed white lead, of great value.

LITHOPONE.

Lithopone is a chemically precipitated pigment containing approximately 70 per cent. barium sulphate, 30 per cent. zinc sulphide, with occasionally occluded impurities, such as zinc oxide, barium chloride, and barium carbonate.

The following method after repeated trials by the authors is considered best for the analysis of this pigment.

Moisture.—Heat 2 grams of the sample for two hours at 105° C. The loss will be moisture. This should be less than one-half of 1 per cent.

Barium Sulphate.—Treat 1 gram of the sample with 10 c.c. of concentrated hydrochloric acid. Add 5 c.c. of bromine water. Evaporate the solution to one-half its volume on a water-bath. Next add 100 c.c. of water and a few drops of dilute sulphuric acid. Boil, filter, wash, and weigh the insoluble residue which should show only the presence of barium sulphate. Examine the insoluble residue for silica and aluminum. Examine the residue as before stated.

Total Zinc.—Determine the total zinc in the filtrate, either gravimetrically or volumetrically, as given under Zinc Oxide on page 1.

Zinc Sulphide.—One gram of the sample is digested at room temperature with 100 c.c. of 1 per cent. acetic acid for one hour. Filter and wash the insoluble residue. Transfer this residue to a beaker, boil with dilute hydrochloric acid, and titrate the zinc directly, as given under Zinc Oxide. The soluble zinc may also be determined in the filtrate by one of the methods, as outlined. Zinc soluble in acetic acid is reported as oxide; zinc insoluble, as zinc sulphide. The filtrate from the acetic acid treatment, after precipitating the zinc as zinc sulphide and subsequent removal, should be examined for barium which might be present as carbonate, and calcium, present as sulphate or carbonate. The presence of barium carbonate is unusual. The average amount of zinc present as zinc oxide in pure lithopone will usually be below 2.5 per cent., though sometimes as high as 5 per cent.

Soluble Salts.—Digest a 2-gram sample with 50 c.c. hot water and test the filtrate for soluble impurities.

BARYTES AND BLANC FINE.

These pigments are the natural and chemically precipitated forms, respectively, of barium sulphate. They vary in their percentage of barium sulphate, according to their purity. The content of barium sulphate should not be less than 95 per cent. The method as used for the analysis of these pigments is as follows:

Moisture.—Heat 2 grams of the sample at 105° C. for two hours. The loss will be moisture.

Loss on Ignition.—Ignite 1 gram of the sample for one-half hour. The loss will be organic matter, free and combined water, and carbon dioxide from any carbonate present. This loss should be reported as loss on ignition.

Barium Sulphate.—Treat 1 gram of the sample with 20 c.c. of dilute hydrochloric acid. Boil, evaporate to dryness, moisten with hydrochoric acid, and take up with water. Boil, filter, and wash. Should lead be present in the insoluble residue, as shown by the action of hydrogen sulphide, treat the insoluble residue with a little 1:1 hydrochloric acid and several drops of sulphuric acid. Filter, wash, and weigh the insoluble residue. Should lead be absent, the last treatment may be omitted. Treat the insoluble residue after washing with an excess of hydrofluoric acid and a few drops of sulphuric acid. Evaporate to dryness, ignite, and determine any loss. This loss will be silica. Examine the filtrate for aluminum oxide, iron oxide, and calcium oxide in the usual way.

Soluble Sulphate.—Soluble sulphates should be determined by treating 1 gram of the sample with 20 c.c. of dilute hydrochloric acid. Dilute with 200 c.c. of hot water, boil, filter, wash, and determine the sulphate with barium chloride in the usual way. Calculate to calcium sulphate. If carbonates are present, as shown by effervescence with hydrochloric acid, calculate the remaining calcium oxide to carbonate. If absent, report as calcium oxide.

WHITING AND PARIS WHITE.

These pigments are the natural and artificial forms, respectively, of calcium carbonate. Dissolve one gram in hydrochloric acid, filter, determine the calcium in the filtrate, as stated, under Gypsum. Determine the carbonic acid by the author's short method, as outlined under the estimation of CO_2. Determine the impurities as usual.

GYPSUM.

This pigment comes to the paint trade in either the hydrated or burnt variety. The amount of water contained in this pigment should therefore be carefully determined.

Total Water.—Ignite 1 gram of the sample to constant weight. The loss will be free and combined water, if carbonates are absent.

Moisture.—Heat 2 grams of the sample at 105° C. for two hours. The loss will be uncombined water.

Calcium.—Calcium is determined by one of the following methods:

Gravimetric Method.—Treat one gram of the sample with dilute hydrochloric acid. Add 150 c.c. of water. Boil, filter, and determine the insoluble. Precipitate the aluminum oxide and iron oxide in the filtrate with ammonium hydroxide and determine in the usual way. The slightly ammoniacal solution is heated to boiling. A boiling solution of 25 c.c. saturated ammonium oxalate is added and the solution allowed to stand from one to two hours or boiled continuously for one-half hour. Filter off the precipitated calcium oxalate, wash with boiling water, and ignite for fifteen or twenty minutes with a Bunsen burner to constant weight. Weigh as calcium oxide and calculate to calcium sulphate. Determine the magnesium in the filtrate in the usual way.

Volumetric Method.—The method as outlined by Treadwell and Hall* is excellent. "The calcium is precipitated in the form of its oxalate, filtered and washed with hot water. The still moist precipitate is transferred to a beaker by means of a stream of water from the wash-bottle, and the part remaining on the filter is removed by allowing warm dilute sulphuric acid to pass through it several times. To the turbid solution in the beaker, 20 c.c. of sulphuric acid 1 : 1 are added and after dilution with hot water to a volume of from 300 to 400 c.c., the oxalic acid is titrated with N/10 potassium permanganate solution. One c.c. of N/10 $KMnO_4 \doteq 0.0020$ gm. Ca."

The sulphuric-acid content is determined by dissolving 1 gram of the sample in concentrated hydrochloric acid, precipitating and weighing as barium sulphate, in the usual way.

Silica Pigments: Silex, Asbestine and China Clay.—The use of the microscope in making a rough examination of these pigments is very valuable, as they all show different characteristics, the asbestine being long and rod-like in its particle shape, the china clay tabloid in shape, and the silica in small, sharp particles.

Moisture.—Heat 2 grams of the sample at 105° C. for two hours. The loss will be moisture.

Loss on Ignition.—Ignite 2 grams of the sample in a platinum crucible for one hour. The loss will be water, unless a large amount of carbonate is present.

Complete Analysis.—One-half a gram of the sample is thoroughly mixed with 10 grams of sodium carbonate, and 1/2 gram of potassium nitrate, in a platinum crucible, and fused until clear. Cool, dissolve in water in a casserole. Make acid with hydrochloric acid, having the casserole covered with a watch glass. Carefully clean out the crucible with a little acid, adding this acid to the main solution. After effervescence has ceased, evaporate to dryness on the sand bath. Cool, moisten the residue with a few

* Treadwell and Hall, analytical chemistry, Vol. II, page 491.

drops of hydrochloric acid, and repeat the evaporation to dryness. Add a few c.c. of concentrated hydrochloric acid, allow to stand for a few minutes, and add about 10 c.c. of hot water. Filter, ignite, and weight as silica. This residue, after weighing, should be treated with an excess of hydrofluoric acid and a few drops of sulphuric acid. Evaporate to dryness and ignite. The loss will be silica. Any residue which remains in the crucible should be again fused with sodium carbonate and the fusion added to the original filtrate. If barytes is found to be present, the original sodium carbonate fusion is dissolved in hot water and the barium carbonate filtered off, dissolved in hydrochloric acid, and precipitated as barium sulphate in the usual way. This filtrate is then added to the original solution and the silica then dehydrated. The iron oxide and aluminum oxide are precipitated in the filtrate with ammonium hydroxide, washed and weighed in the usual way. The precipitate of iron and aluminum oxides is fused in a platinum crucible with potassium acid sulphate solution, the fusion taken up with water, treated with concentrated sulphuric acid, and the iron titrated, after being reduced with zinc, by means of potassium permanganate. The difference between this iron oxide and the combined weight of iron and aluminum oxide is the alumina. The filtrate from the iron and aluminum-precipitation is treated with ammonium oxalate, and the calcium determined as described under gypsum. The filtrate from the calcium is tested for magnesium with sodium hydrogen phosphate, and determined in the usual way. Carbon dioxide present is determined in a separate sample, as before given. The amount found is calculated to calcium carbonate. Any excess of calcium is reported as oxide. The magnesium is calculated as magnesium oxide, unless the carbon dioxide is in excess of the calcium present, in which case it is calculated to magnesium carbonate, and the remainder of the magnesium to the oxide.

Sodium and Potassium.—Mix 1 gram of the sample with one part of ammonium chloride and eight parts of pure calcium carbonate. Heat to dull redness and thus convert the sodium and potassium to chlorides. Cool, take up with water. Filter and precipitate the calcium with ammonia and ammonium oxalate, and again filter and wash. Evaporate to dryness in a weighed platinum dish, on the water bath. Take care to avoid spattering, as dryness is reached. Finally heat with a Bunsen burner to a very faint, red heat. Cool and weigh. This weight represents the total potassium and sodium as chlorides. Take up the residue in water and add an excess of platinic chloride solution. Evaporate to small volume, take up with 80 per cent. alcohol, filter in a Gooch crucible, wash with alcohol. Dry in the steam oven. Calculate the potassium in the potassium platinic chloride to potassium oxide and then to potassium chloride. The difference between the potassium chloride and the combined weight of the chlorides is the sodium chloride which may then be calculated to sodium oxide. It is well to convert the chlorides of sodium and potassium into sulphate, after the original chloride solution has reached a small volume, by the addition of a few c.c. of sulphuric acid. The sulphates of these metals are less volatile than the chlorides, and greater accuracy can be obtained by this method. After weighing, the potassium is determined as before stated.

OCHERS, UMBERS, AND SIENNAS.

Mannhardt's* method for the determination of these pigments is an excellent one, and follows:

"One gram of pigment is treated with 20 c.c. of 1 to 1 hydrochloric acid on the steam plate in a covered beaker. Small amounts of stannous chloride crystals are gradually introduced into the hot liquid until the solution and insoluble are nearly colorless and the iron is all reduced to the ferrous state. (In umbers, which all contain manganese, this should be removed as $MnO_2.2H_2O$ by boiling with excess of ammonium hypobromite in ammoniacal solution before taking up the reduction with stannous chloride.) The ferrous and stannous chloride solution is now stirred into 10 c.c. of 6 per cent. solution of mercuric chloride to remove the stannous salt. An equal bulk of strong hydrochloric acid is then added and the solution titrated with 3 N/10 bichromate using a pale yellow solution of ferricyanide as external indicator. Nearing the end point, more acid is added, to bring the solution to an olive color. One c.c. 3 N/10 equals .024 grams Fe_2O_3. Raw siennas may contain some ferrous iron. This is determined by dissolving 1 gram of pigment and 1 gram of bicarbonate of soda with hydrochloric acid,

* Hans Mannhardt, Select Methods of Paint Analysis, pp. 22 and 23.

using a small flask provided with a Bunsen valve. The ferrous iron is then titrated directly. One c.c. 3 N/10 equals .0216 FeO.

"The hydrated peroxide of manganese is best determined as manganous pyrophosphate Fresenius Quant., Vol. I, page 297.

"The complete analysis of an ocher, umber, or sienna requires the fusion method of solution. In the interpretation of results any alumina is calculated to clay and the excess silica returned as such. The loss on ignition is determined on a separate sample and is made up of moisture (organic matter sometimes), the combined water of the clay, and the combined water of the limonite $2Fe_2O_3.3H_2O$. The Mn_3O_4 of the raw umber is probably also hydrated. Burnt ochers, umbers, and siennas of course carry very little free and combined water."

IRON OXIDES.

(Venetian Red, Metallic Brown, Indian Red.)

Test for free sulphuric acid, by boiling in water, filtering off and using litmus paper. Estimate the amount by titration.

Walker's* method for the analysis of oxide and earth pigments, which also includes the analysis of ochers, umbers, and siennas, is as follows:

"Iron oxide is extensively used as a paint. The native oxide naturally varies very much in its composition. In general, however, only the poorer grades of native hematite are used as paints. Artificial iron oxide pigments, made by calcining copperas, may be practically pure ferric oxide. Umbers, ochers, and siennas are earthy substances containing iron and manganese oxides and more or less organic matter.

"The methods of analysis are very much the same as those for iron ores. It is generally sufficient to determine moisture, loss on ignition, insoluble in hydrochloric acid, ferric oxide, and manganese dioxide. If much organic matter is present, roast 2.5 grams in a porcelain dish at a low temperature until all organic matter is destroyed, add 25 c.c. of hydrochloric acid, cover with a watch glass, and heat on the steam bath for two hours; then add 10 c.c. of sulphuric acid (1:1), evaporate, and

* P. H. Walker. Bulletin No. 109, revised, Bureau of Chemistry, U. S. Dept. of Agriculture, pp. 33 and 34.

heat until fumes of sulphur trioxide are evolved and all the hydrochloric acid is driven off. Cool, add 50 c.c. of water, boil until all of the iron sulphate is dissolved, filter into a 250 c.c. graduated flask, fill to the mark, mix, and take out a portion of 50 c.c. for iron determinations. For the determination of iron reduce with zinc, titrate with standard potassium permanganate, and calculate to ferric oxide. For the determination of manganese transfer 100 c.c. of the solution (corresponding to 1 gram) to a 500 c.c. flask, add sodium hydroxide solution until nearly but not quite neutral. Then add an emulsion of zinc oxide in water in slight excess, shake until all of the iron is thrown down, fill to the mark, mix, let settle, filter through a dry paper, and use portions of 100 c.c. (corresponding to 0.2 gram) for the manganese determination. Transfer to a 300 c.c. Erlenmeyer flask, heat to boiling, titrate with standard permanganate. The flask must be shaken during the titration and some experience is necessary to determine the end point, which is best seen by looking through the upper layer of liquid and observing when the pink tinge from the permanganate does not disappear on shaking. A standard potassium permanganate solution, the iron factor of which is known, may be used to determine manganese. The iron factor multiplied by 0.2952 gives the manganese factor. In some cases this method of attack will not separate all of the iron from the insoluble matter. In such a case the insoluble must be fused with a mixture of sodium and potassium carbonates, dissolved in water, evaporated with excess of sulphuric acid, filtered from the insoluble, and this solution added to the first one.

"Another method is as follows: Roast 5 grams of the powder, digest with 25 c.c. of hydrochloric acid, evaporate to dryness, moisten with hydrochloric acid, dissolve in water, filter, and wash the residue; ignite the residue in a platinum crucible, add sulphuric acid and hydrofluoric acid, drive off the latter, and heat until copious fumes of sulphuric anhydride come off. Add potassium hydrogen sulphate and fuse, dissolve in water, filter from any remaining insoluble (barium sulphate), unite the two solutions, make up to 500 c.c. and use aliquots for the iron and manganese determinations. For the determination of iron place 100 c.c. in a flask, add about 3 grams of zinc, put a funnel into the neck of the flask, heat when the action slackens; if basic salts separate out add a few drops of hydrochloric acid. When all of the iron is reduced, add 30 c.c. of sulphuric acid (1:2), and as soon as all of the zinc is dissolved and the solution is cool, titrate with potassium permanganate.

"For the determination of manganese use 50 c.c., evaporate to a very small bulk, add strong nitric acid and evaporate the hydrochloric acid; add 75 c.c. of strong nitric acid, which should be free from nitrous acid, and 5 grams of potassium chlorate, heat to boiling and boil fifteen minutes; then add 50 c.c. of nitric acid and more potassium chlorate. Boil until yellow fumes cease to come off, cool in ice water, filter on asbestos, and wash with colorless, strong nitric acid; suck dry and wash out remaining nitric acid with water, transfer the precipitate and the asbestos to a beaker, add a measured excess of standard solution of ferrous sulphate in dilute sulphuric acid, stir until all of the manganese dioxide is dissolved, and titrate the remaining ferrous sulphate with potassium permanganate. A ferrous solution of about the proper strength is made by dissolving 10 grams of crystallized ferrous sulphate in 900 c.c. of water and 100 c.c. of sulphuric acid (specific gravity 1.84). This solution is titrated with standard potassium permanganate. The reaction taking place when the manganese dioxide acts on the ferrous sulphate is $MnO_2+2FeSO_4+2H_2SO_4=MnSO_4+Fe_2(SO_4)_3+H_2O$. Hence the iron value of permanganate multiplied by 0.491 gives the value in manganese."

RED LEAD AND ORANGE MINERAL.

"These pigments in the pure state are oxides of lead (approximately Pb_3O_4), being probably mixtures of compounds of varying proportions of lead monoxide and lead dioxide. Moisture, insoluble impurities, and total lead may be determined by the methods given under chrome yellow; or, in the absence of alkaline earth metals, the lead may be determined as sulphate in nitric acid solution (dissolve by adding a few drops of hydrogen dioxide) by evaporating with an excess of sulphuric acid until fumes of sulphuric anhydride are evolved. Determine as sulphate in the usual way.

"The lead dioxide (PbO_2) may be determined as follows: Weigh 0.5 gram of the very finely ground pigment into a 150 c.c.

Erlenmeyer flask. Mix in a small beaker 15 grams of crystallized sodium acetate, 1.2 grams of potassium iodide, 5 c.c. of water, and 5 c.c. of 50 per cent. acetic acid. Stir until all is liquid, pour into the Erlenmeyer flask containing the lead, and rub with a glass rod until all of the lead is dissolved; add 15 c.c. of water, and titrate with tenth-normal sodium thiosulphate, using starch as indicator. A small amount of lead may escape solution at first, but when the titration is nearly complete this may be dissolved by stirring. The reagents should be mixed in the order given, and the titration should be carried out as soon as the lead is in solution, as otherwise there is danger of loss of iodine. One cubic centimeter of tenth-normal sodium thiosulphate corresponds to 0.011945 gram of lead dioxide, or 0.034235 gram of red lead (Pb_3O_4).

"These colored lead pigments may have their color modified by the addition of organic coloring matters. As a general rule, such adulteration may be detected by adding 20 c.c. of 95 per cent. alcohol to 2 grams of the pigment, heating to a boil, and allowing to settle. Pour off the alcohol, boil with water, and allow to settle, then use very dilute ammonium hydroxide. If either the alcohol, water, or ammonium hydroxide is colored, it indicates organic coloring matter. The quantitative determination of such adulteration is difficult and must generally be estimated by difference."

Vermilion.

"True vermilion, or, as it is generally called, English vermilion, is sulphide of mercury. On account of its cost it is rarely used in paints, and is liable to gross adulteration. It should show no bleeding on boiling with alcohol and water and no free sulphur by extraction with carbon disulphide. A small quantity mixed with five or six times its weight of dry sodium carbonate and heated in a tube should show globules of mercury on the cooler portion of the tube. The best test for purity is the ash, which should be not more than one-half of 1 per cent. Make the determination in a porcelain dish or crucible, using 2 grams of the sample. Ash in a muffle or in a hood with a very good draft, as the mercury fumes are very poisonous. It is seldom necessary to make a determination of the mercury; but if this is required, it may be determined by mixing 0.2 gram of the vermilion with 0.1 gram of very fine iron filings, or better "iron by hydrogen." Mix in a porcelain crucible and cover with a layer 10 mm. thick of the iron filings, place the crucible in a hole in an asbestos sheet so that it goes about half way through, cover with a weighed, well fitting, gold lid which is hollow at the top, fill this cavity with water, heat the crucible for fifteen minutes with a small flame, keep the cover filled with water, cool, remove the cover, dry for three minutes at 100° C., and thirty minutes in a desiccator, and weigh. The increase in weight is due to mercury. The mercury can be driven off from the gold by heating to about 450° C. A silver lid may be used, but gold is much better.

"Another method is to place in the closed end of a combustion tube, 45 cm. long and 10 to 15 mm. in diameter, a layer of 25 to 50 mm. of roughly pulverized magnesite, then a mixture of 10 to 15 grams of the vermilion with four or five times its weight of lime, followed by 5 cm. of lime, and plug the tube with asbestos. Draw out the end of the tube and bend it over at an angle of about 60°. Tap the tube so as to make a channel along the top, and place it in a combustion furnace with the bent neck down, resting with its end a little below some water in a small flask or beaker. Heat first the lime layer, and carry the heat back to the mixture of lime and pigment. When all the mercury has been driven off, heat the magnesite, and the evolved carbon dioxide will drive out the last of the mercury vapors. Collect the mercury in a globule, wash, dry, and weigh.

"Genuine vermilion is at the present time little used in paints. Organic lakes are used for most of the brilliant red, scarlet, and vermilion shades. These organic coloring matters are sometimes precipitated on red lead, orange mineral, or zinc oxide; but as a usual thing the base is barytes, whiting, or china clay. Paranitraniline red, a compound of diazotized paranitraniline and beta-naphthol is largely employed; but a number of colors may be used. To test for red colors in such a lake the following method from Hall* may be of value, though other colors may be employed, which makes the table of only limited use.

* "The Chemistry of Paints and Paint Vehicles," p. 29.

Table V.—Determination of Red Colors in Organic Lake.

Reagent.	Source of color.				
	Alizarine.	Eosin.	Para-nitraniline.	Ortho-anisidine.	Scarlet (2R).
Sulphuric acid, conc.	Dark brown with reddish undertone becoming light yellow on diluting.	Changes to yellow; light red solution with excess of sodium hydroxide.	Purple; color returning on diluting.	Purple; red on diluting.	Darkened; lighter on diluting.
Hydrochloric acid, conc.	Color becomes "muddy."	Changes to yellow; reddish light solution with excess of sodium hydroxide.	Color slightly darkened; lighter on diluting.	Dark purplish red; lighter on diluting.	Darkened; lighter on diluting.
Sodium hydroxide conc. sol.	Dark reddish-brown; little change on diluting.	Little change; light solution on diluting.	Color changed to brownish-red; light red on diluting.	Little change.	Reddish solution on diluting.
Alcohol	Insoluble.......	Reddish fluorescent solution.	Slight yellowish, orange solution.	....	Slight reddish solution.
Sodium hydroxide conc. and alcohol.	Purplish; dark brown on diluting.	....	Purple; color returning on diluting.	Little change.	Color darker; reddish solution on diluting.

"It is well also to try the action of reducing and oxidizing agents such as stannous chloride, ferric chloride, etc. (See Zerr, Bestimmung von Teerfarbstoffen in Farblacken; also Schultz and Julius, A Systematic Survey of the Organic Coloring Matters.)

"Paranitraniline red is soluble in chloroform. It is also well to try the solvent action on different reds, of sodium carbonate, etc. The amount of organic pigment present in such reds is generally very small, and when it cannot be determined by ignition owing to the presence of lead, zinc, or carbonate, it is best determined by difference."*

BLUE PIGMENTS.

The only two blue pigments which come into great use in the paint trade are Ultramarine and Prussian blue.

Ultramarine Blue.

Ultramarine blue is formed by burning together silicates, aluminates, sulphur, charcoal, and soda. Although of indefinite composition, it is essentially a silicate and sulphide of sodium and aluminum. This compound may be identified by its evolution of hydrogen sulphide when treated with acid. Although seldom used in the paint trade, if an analysis is desired, it may be carried out in the following way.

Moisture.—Heat 2 grams of the sample at 105° C. for two hours. The loss will be moisture.

Silica.—One gram of the sample is digested with 20 c.c. dilute hydrochloric acid to complete decomposition, taking care to avoid spattering. Evaporate to dryness, dehydrate, cool, moisten with a few drops of concentrated hydrochloric acid. Repeat the evaporation and dehydration, cool, moisten with a few c.c. concentrated hydrochloric acid, dilute with hot water, filter, and weigh the insoluble residue. Treat the contents of the crucible with an excess of hydrofluoric acid and a few drops of sulphuric acid, evaporate to dryness, and ignite. The loss will be silica. Any residue remaining after this treatment should be examined.

Alumina.—The filtrate from the silica is treated with ammo-

* Bulletin 109, revised, Bureau of Chemistry, U. S. Dept. of Agriculture, pp. 31, 32 and 33.

nium hydroxide until faintly alkaline. The aluminum hydroxide precipitated is filtered, washed, and weighed as Al_2O_3. If iron is present it may be separated from alumina by the fusion method as stated under the analysis of white pigments.

Sodium Oxide.—Treat the filtrate after the removal of the aluminum with sulphuric acid until faintly acid. Evaporate to dryness, weigh the sodium sulphate in a weighed dish. Should calcium be present, the sodium is determined in the filtrate after the removal and determination of the calcium as oxalate, in the way previously stated.

Total Sulphur.—One gram of the sample is fused with a mixture of potassium nitrate and sodium carbonate. Dissolve the fused mass in hydrochloric acid and boil, after the addition of concentrated nitric acid. Filter off the insoluble residue. Determine the sulphate in the filtrate in the usual way, by precipitation as barium sulphate with barium chloride. The sodium peroxide method for the oxidation of sulphur is often used for this determination.

Sulphur Present as Sulphate.—One gram of the sample is boiled with dilute hydrochloric acid until all the hydrogen sulphide has been driven off. Filter and wash the insoluble residue with hot water, precipitate the sulphur present in the filtrate, as usual. The amount of the sulphate present is deducted from the total sulphur found as sulphate by the fusion method, —the difference calculated to sulphur and reported as such.

Ultramarine blues usually vary between the following limits:

Sulphur,	10 to 14 per cent.
Sulphur trioxide,	2 to 5 per cent.
Alumina,	20 to 27 per cent.
Silica,	39 to 45 per cent.
Soda,	15 to 20 per cent.

Prussian Blue (Antwerp Blue and Chinese Blue*).

This pigment is a double iron and potassium salt of hydroferricyanic and hydroferrocyanic acids. Its composition is rather indefinite and consequently analyses of the compound give varied results. The following method of analysis has been much used by the authors:

* Chinese Blue often contains a small percentage of tin salts. They should be looked for in the qualitative examination.

Moisture.—Two grams of the sample are heated at 105° C. for two hours. The loss will be moisture. Less than 7 per cent. moisture should be present in dry Prussian blue.

Iron and Aluminum Oxides.—One gram of the sample is ignited at a low temperature until the last trace of blue has been decomposed. The ignition must be low, so as to prevent any iron from being rendered difficultly soluble in hydrochloric acid. After cooling, treat the mass with 25 c.c. hydrochloric acid 1 : 1 and digest for one hour. No residue will remain if a pure Prussian blue is being examined. In some instances a residue does remain after this treatment, due to the presence of inert bases upon which the blue has been precipitated, such as barytes. Should such an insoluble residue be present, evaporate the solution to dryness, take up with hot water and dilute hydrochloric acid, boil, filter, wash, and weigh the insoluble residue. Examine this for silica, barium sulphate, and alumina.

The filtrate from the last treatment, or the original solution, should no insoluble matter be present, is divided into aliquot portions. One of these portions is made faintly acid with ammonium hydroxide and the combined aluminum and iron hydroxides formed filtered, washed, and weighed in the usual way. The iron may be determined in the combined oxides by fusion with potassium acid sulphate and titration with potassium permanganate in the usual way. The filtrate should be examined for calcium present as calcium sulphate.

Nitrogen.—Nitrogen is determined by the Kjehdahl-Gunning method.

Commercial Method.—According to Parry and Coste,* the percentage of Prussian blue can be determined with sufficient accuracy for commercial purposes by multiplying the percentage of nitrogen by 4.4 and the percentage of iron by 3.03. The iron in this case may be titrated directly after solution in hydrochloric acid.

Akalies and Sulphuric Acid.—Alkaline salts are estimated in the filtrate after the removal of calcium, in the usual way One of them only will usually be present. Calculate to metal, then to sulphate, accounting for all the sulphuric acid as alkaline sulphate.

Sulphuric acid is determined in the other aliquot portion, by

* The Analyst, 1896, 21, 225 to 230.

precipitation as barium sulphate, in the usual way. It is calculated to alkaline sulphate.

Yellow and Orange Pigments (American Vermilion, Chrome Yellow, Lemon Chrome, and Orange Chrome).

The above pigments, varying in color, always contain chromates. The light colored pigments usually contain lead sulphate or other insoluble lead compounds. The darker pigments sometimes contain basic lead chromate. It is advisable to treat the pigment with alcohol to determine the presence of any organic coloring matter. Pure chrome yellow should contain only lead chromate and insoluble lead compounds. These pigments are analyzed in the following way:

Moisture.—Heat 2 grams of the sample at 105° C. for two hours. Loss will be moisture.

Insoluble Residue.—One gram of the sample is treated with 25 c.c. concentrated hydrochloric acid, the solution boiled, during which a few drops of alcohol are added one at a time. Dilute to 100 c.c. with hot water, continue the boiling for ten minutes, filter, wash, weigh the insoluble residue. This residue is examined for silica, barium sulphate, and alumina.

Lead.—After neutralizing the greater portion of the acid present with ammonium hydroxide, after the removal of the insoluble residue, the solution is diluted to about 300 c.c. with water. Precipitate the lead completely with hydrogen sulphide, allow to settle, evaporate, wash with hydrogen sulphide water. Dissolve the lead sulphide in hot dilute nitric acid, add an excess of sulphuric acid, evaporate until heavy fumes of sulphuric acid are given off, cool, dilute with water, add an equal volume of alcohol, filter, determine the lead either volumetrically or gravimetrically, as given under White Lead.

Chromium.—The filtrate from the lead precipitate is boiled until all the hydrogen sulphide is expelled. Precipitate the chromium as chromium hydroxide with ammonium hydroxide, observing the usual precautions for preventing the precipitation of any zinc. Filter, wash, and weigh as chromic oxide. Calculate to chromic anhydride.

Zinc.—Precipitate the zinc in the filtrate from the chromium precipitation, with hydrogen sulphide, as given under Zinc

Oxide, filter, wash, and weigh. Determine the zinc either gravimetrically or volumetrically, as before outlined.

Calcium and Magnesium.—Determine the calcium and magnesium in the filtrate in the usual way.

Sulphuric Acid.—The combined sulphate is estimated by dissolving 1 gram of the sample in hydrochloric acid, removing the insoluble residue by filtration, precipitating as barium sulphate, in the usual way. The barium sulphate precipitated will be pure, providing the solution is kept dilute and hot, otherwise lead sulphate may be precipitated. Should the precipitate become contaminated, the sulphate is best determined by the Hughes method, as outlined on page 17.

GREEN PIGMENTS.

(Chrome Green.)

The green pigments which are of most importance are those consisting of a mixture of Prussian blue and chrome yellow. Organic colors are sometimes used to give the pigment a bright tint. It is well to examine all green pigments for coloring matter by boiling with alcohol.

Owing to the various composition of greens, due to the method of manufacture, a chemical examination is attended with many difficulties, and in many cases is of small value in determining the true value of the green. The usual methods of color assay for strength, by mixing with a weighed portion of white-base pigment and observing the tint produced, should be made. The pigment should also be carefully examined by means of the microscope, so as to determine whether the green is a product made by precipitating the two pigments together or by mixing the blue and yellow pigments after separate precipitation. The former have the greater value. A good green will show the presence of green and blue particles but no yellow, while a poor green will show yellow and blue particles mixed with green. Analysis of the green may be made in the following way:

Moisture.—Heat 2 grams of the sample at 105° C. for two hours. The loss will be moisture.

Insoluble Residue.—One gram of the sample is heated at a very low heat in a casserole until the blue color has been completely destroyed, keeping the temperature sufficiently low so as not to render any of the iron or lead chromates insoluble. Cool, add 30 c.c. of concentrated hydrochloric acid, boil until all the soluble constituents have passed into solution. This can be hastened by the addition of a few drops of alcohol added one at a time. Dilute with water, boil, filter, wash, and weigh the insoluble residue. This residue is then carefully examined for silica, barytes, and occasionally alumina in the manner given under the analysis of a white pigment.

Lead.—The excess of acid present in the filtrate from the insoluble residue is neutralized with ammonia until barely acid. Dilute to 300 c.c., allow to cool, and treat with hydrogen sulphide until the lead sulphide is completely precipitated. Allow the precipitate to settle, filter, wash, dissolve in nitric acid, determine the lead as described under the analysis of yellow pigments.

Iron, Alumina, and Chromium.—The filtrate from the lead precipitate is boiled until all the hydrogen sulphide has been expelled. Add a few drops of nitric acid, boil for a few minutes to complete oxidation of the iron, precipitate the iron, alumina, and chromium as hydroxides with ammonium hydroxide. This precipitate is filtered off, washed, and dissolved in hydrochloric acid and made up to a definite volume. In one portion the iron, aluminum, and chromium hydroxides are precipitated with ammonium hydroxide, filtered, washed, and weighed together. Another portion is treated in a flask with an excess of potassium hydroxide and bromine water until the iron hydroxide has assumed its characteristic reddish brown color. Dilute with water, filter, and wash. Dissolve the iron hydroxide in hydrochloric acid and determine the iron in the solution either volumetrically or gravimetrically, as stated under the analysis of iron oxides. The filtrate from the iron precipitate is made acid with nitric acid, and the alumina precipitated and determined in the usual way by precipitation with ammonium hydroxide. Chromium is determined in the filtrate by reduction to a chromic salt with hydrochloric acid and alcohol, precipitated with ammonium hydroxide and weighed as oxide.

Calcium and Magnesium.—Calcium and magnesium are determined in the filtrate from the iron, alumina, and chromium precipitation.

Sulphuric Acid.—One gram of the sample, after ignition until the blue is completely decomposed, as before stated, is dissolved in 30 c.c. of concentrated hydrochloric acid, diluted with water, boiled, filtered, and washed. Determine the barium sulphate in the usual way.

Nitrogen.—Nitrogen is determined as stated under the analysis of Prussian blue.

Interpretations of Results.—The Prussian blue present is determined by multiplying the iron found by 3.03 or the nitrogen found by 4.4. The sulphate is calculated to lead sulphate and calcium sulphate, should calcium be present, and the chromium to lead chromate.

BLACK PIGMENTS (Bone Black, Drop Black, Ivory Black, Lampblack, Mineral Black, Graphite).

The usual black pigments are those consisting of carbon, existing either in the crystalline form (graphite) or the amorphous form which includes all the many artificial carbon blacks. Many paints are made at the present time from coal-tar and asphaltic mixtures, and it is deemed advisable to treat of these, owing to many difficulties attending their analysis, under a separate chapter. For the analysis of simple black pigments the following method will be found to embrace all the essential points needed for their examination. An extraction of the black pigment should be made with ether to determine the presence of oil. If present, it may be determined gravimetrically in a fat extraction apparatus.

Moisture.—Dry 2 grams of the sample at 105° C. for two hours. Loss will be moisture.

Volatile Matter.—One gram of the sample is heated in a large covered crucible with a Bunsen burner for ten minutes. The loss will be volatile matter.

Ash.—Two grams of the sample are ignited at a bright red heat until the carbon has all been driven off. The residue will be ash. Should the black pigment consist of graphite, the ignition must be made with the aid of oxygen. Ignite until all the carbon is burned off and report as carbon. Should carbonates be present, the ash is mixed with a small amount of ammonium carbonate and again ignited and weighed. This will reconvert any carbonates which may have been converted to oxides on ignition.

Soluble Salts.—The ash from the above is boiled with concentrated hydrochloric acid, diluted with water, again boiled, filtered, and the insoluble residue weighed. Should it be desired, the insoluble residue may be examined for silica, barium sulphate, and alumina. The filtrate is examined for phosphoric acid, calcium, and magnesium.

CHAPTER II.

THE ANALYSIS OF MIXED PIGMENTS AND PAINTS.

After making the usual observations as to the statements on the label of the can, gross weight of package, etc., the can is opened, and the material well stirred, preferably by transferring the entire contents to a large receptacle. A uniform sample is then withdrawn for analysis. The original can may be calibrated by pouring in a standard volume of water.

A portion (about 15 grams) of the sample of paint withdrawn for analysis is weighed into a 2-ounce tared centrifuge bottle, and 40 c.c. of a vehicle solvent* added. The bottle is then capped and placed in the centrifuge machine for twenty minutes, when it is removed and the vehicle poured off from the settled pigment. Successive portions of solvent are again added, and the operation repeated several times until the pigment has been thoroughly extracted. The bottle and its contents of pigment is dried at 105° C. in an air-bath, and, after weighing, the loss is calculated to the percentage of vehicle in the paint, the balance being, of course, pigment.

The dried pigment is now ready for analysis.

GENERAL METHODS FOR THE ANALYSIS OF A MIXED WHITE PAINT.

Any of the constituents mentioned under the analysis of simple white pigments may be present in a mixed white paint. The analysis may be considerably shortened by a preliminary qualitative examination. It is assumed, however, that zinc sulphide will not be present with lead compounds, as such a mixture is apt to blacken. Either of the following general methods may be used in such an analysis.

Method 1.—One gram of the dry pigment is treated with 20 c.c. (1 : 1) hydrochloric acid. Evaporate to dryness, moisten with a few cubic centimeters of concentrated hydrochloric acid,

* 50 parts benzol, 40 parts wood alcohol, 10 parts acetone.

allow to stand for several minutes, dilute to 100 c.c. with hot water, boil, filter, and wash the insoluble residue with hot water. It is advisable to treat the residue after washing into the original beaker with 1 : 1 hydrochloric acid and 2 c.c. of dilute sulphuric acid. Boil, filter and wash. This will usually remove the last traces of lead. Should no barium sulphate be present, the hydrochloric acid-sulphuric acid treatment may be used in the first instance. The insoluble residue is ignited and weighed. After weighing, treat the contents of the crucible with an excess of hydrofluoric acid and a few drops of sulphuric acid. Evaporate to dryness and ignite. The loss will be silica.

The residue, after the silica has been removed by volatilization, is fused with potassium acid sulphate, and the melt taken up in hot water. Filter, wash, and weigh, as given under Barium Sulphate. The filtrates from the original fusion and extraction are then combined. A sodium carbonate fusion may also be resorted to in place of potassium acid sulphate at this point.

Warm the combined filtrates and pass in hydrogen sulphide until the lead is completely precipitated, having from 3 to 4 1/2 c.c. of concentrated hydrochloric acid for every 100 c.c. of neutral solution. If the color of this precipitate is gray, it is a sign that some zinc is being precipitated, and acid must be added. Should the precipitate be reddish-black, the solution is too acid. Cool the contents of the beaker, filter, and wash. The lead is estimated as described under Basic Carbonate White Lead. Boil the filtrate from the lead precipitate until the hydrogen sulphide is completely removed. Add a few drops of nitric acid, ammonium chloride, and ammonium hydroxide to faint excess. Filter, wash, and weigh the aluminum oxide and any iron oxide which is precipitated in the usual way, observing the usual precautions to prevent the precipitation of zinc. Should the residue be colored, indicating the presence of iron, the iron oxide may be separated from the alumina by fusion with potassium acid sulphate. Take up the fusion with water, acidify with sulphuric acid, reduce with zinc, and titrate in the usual way with potassium permanganate. Calculate to Fe_2O_3. The difference between this and the total oxides will be Al_2O_3. The amount of iron oxide present in the case of white pigments will be in most cases inappreciable. Acidify the filtrate from the iron oxide and alumina faintly with acetic acid and heat to boiling. Saturate with H_2S and boil for ten minutes. Determine the zinc either gravimetrically or volumetrically, as under Zinc Oxide. To the boiling alkaline filtrate add boiling ammonium oxalate, and determine the calcium in the usual way. Precipitate the magnesium in the filtrate with sodium hydrogen phosphate and determine in the usual way.

Barium carbonate is sometimes found in mixed paints, and it is well to test for this before the precipitation of calcium. A relatively large percentage of magnesium denotes the presence of asbestine.

The calculation of aluminum oxide to clay and determination of the silica present is carried out, according to Scott, as follows:

$$\text{Weight of } Al_2O_3 \times 2.5372 = \text{Weight of clay.}$$
$$\text{Weight of Clay} \times .4667 = \text{Weight of } SiO_2 \text{ in clay.}$$

Any difference greater than 5 per cent. may be considered silica.

Estimation of SO_4 in the lead sulphate which may be present. Treat 1 gram of the sample with 200 c.c. of water and dilute hydrochloric acid. Add several pieces of spongy zinc and boil. After complete precipitation of the lead, filter and wash. Precipitate the sulphuric acid in the usual way, as barium sulphate. Filter, wash, and weigh. Calculate the SO_4 to lead sulphate.

Method 2.—The second method depends upon the separation of the compounds soluble in acetic acid from the insoluble compounds.

Moisture.—Dry 2 grams of the sample at 105° C. for two hours. The loss will be moisture.

Residue Insoluble in Acetic Acid.—One gram of the sample is boiled with a mixture of 10 c.c. 95 per cent. acetic acid and 25 c.c. of water. Filter and wash the insoluble residue. The filtrate is carefully preserved for quantitative examination. The insoluble residue, after washing into a beaker, is treated with 25 c.c. of hydrochloric acid (1.2) and 5 grams of ammonium chloride. Heat on the steam-bath for a few minutes, dilute with hot water to about 300 c.c., boil, filter, wash, and weigh the insoluble residue. This residue is examined for silica, aluminum, and barium sulphate, as stated under the first method for the general analysis of pigments. The lead is precipitated in the filtrate in the usual way with hydrogen sulphide. The lead sulphide precipitate is filtered, washed, dissolved in nitric acid, evaporated in the presence of sulphuric acid, and determined either gravimetrically or volumetrically, as stated under Basic Carbonate-White Lead. The lead found is calculated to lead sulphate. The filtrate from the lead precipitate is examined for aluminum and calcium in the usual way. Any calcium is calculated to calcium sulphate.

Compounds Soluble in Acetic Acid.—The filtrate from the acetic acid treatment may contain lead, zinc, barium, calcium and magnesium. The zinc and lead are removed by passing hydrogen sulphide into the hot solution until completely precipitated. Filter, wash, dissolve in nitric acid, evaporate in the presence of sulphuric acid, and determine the lead, as before stated. Calculate to basic carbonate of lead. The zinc is determined in the filtrate from the lead either gravimetrically or volumetrically, as stated under Zinc Oxide. Calculate to zinc oxide. The filtrate from the lead and zinc precipitate, after removal of the hydrogen sulphide by boiling, is examined for barium, calcium, and magnesium, as stated in the first method for the analysis of white pigments. Calculate to carbonates.

The total sulphate present, both soluble and insoluble, can best be determined by the method as outlined under the analysis of zinc lead and leaded zinc.

Should the qualitative examination of the mixed pigment show definitely its composition, recourse can be had to such shortened methods as given by Thompson.*

"*Sample* 1 is a mixture of barytes, white lead, and zinc oxide.

"Two 1-gram portions are weighed out. One is dissolved in acetic acid and filtered, the insoluble matter ignited and weighed as barytes, the lead in the soluble portion precipitated with bichromate of potash, weighed in Gooch crucible as chromate, and calculated to white lead.

"The other portion is dissolved in dilute nitric acid, sulphuric acid added in excess, evaporation carried to fumes, water added, the zinc sulphate solution filtered from barytes and lead sulphate and precipitated directly as carbonate, filtered, ignited, and weighed as oxide.

*J. Soc. Chem. Ind., June 30, 1896, Vol. XV, No. 6, pp. 433, 434.

"*Sample* 2 is a mixture of barytes and so-called sublimed white lead.

"Weigh out three 1-gram portions. In one determine zinc oxide as in Case 1. Treat a second portion with boiling acetic acid, filter, determine lead in filtrate and calculate to lead oxide. Treat third portion by boiling with acid ammonium acetate, filter, ignite, and weigh residue as barytes, determine total lead in filtrate, deduct from it the lead as oxide, and calculate the remainder to sulphate. Sublimed lead contains no hydrate of lead, and its relative whiteness is probably due to the oxide of lead being combined with the sulphate as basic sulphate. Its analysis should be reported in terms of sulphate of lead, oxide of lead, and oxide of zinc.

"*Sample* 3 is a mixture of barytes, sublimed lead, and white lead.

"Determine barytes, zinc oxide, lead soluble in acetic acid and in ammonium acetate, as in Case 2; also determine carbonic acid, which calculate to white lead, deduct lead in white lead from the lead soluble in acetic acid, and calculate the remainder to lead oxide.

"*Sample* 4 is a mixture of barytes, white lead, and carbonate of lime.

"Determine barytes and lead soluble in acetic acid (white lead) as in Case 1. In filtrate from lead chromate precipitate lime as oxalate, weigh as sulphate, and calculate to carbonate. Chromic acid does not interfere with the precipitation of lime as oxalate from acetic acid solution.

"*Sample* 5 is a mixture of barytes, white lead, zinc oxide, and carbonate of lime.

"Determine barytes and white lead as in Case 1. Dissolve another portion in acetic acid, filter and pass sulphuretted hydrogen through the boiling solution, filter, and precipitate lime in filtrate as oxalate; dissolve mixed sulphides of lead and zinc in dilute nitric acid, evaporate to fumes with sulphuric acid, separate, and determine zinc oxide as in Case 1.

"*Sample* 6 is a mixture of barytes, white lead, sublimed lead, and carbonate of lime.

"Determine barytes, lead soluble in acetic acid and in ammonium acetate, as in Case 2, lime and zinc oxide, as in Case 5, and carbonic acid. Calculate lime to carbonate of lime, deduct carbonic acid in it from total carbonic acid, calculate the remainder of it to white lead, deduct lead in white lead from lead soluble in acetic acid, and calculate the remainder to oxide of lead.

"*Sample* 7 contains sulphate of lime.

"Analysis of paints containing sulphate of lime present peculiar difficulties from its proneness to give up sulphuric acid to lead oxide or white lead if present. Sulphate of lime and white lead boiled in water are more or less mutually decomposed with the formation of sulphate of lead and carbonate of lime. A method for the determination of sulphate of lime is by prolonged washing with water with slight suction in a weighed Gooch crucible. This is exceedingly tedious, but thoroughly accurate. A reservoir containing water may be placed above the crucible, and the water allowed to drop slowly into it. This may take one or two days to bring the sample to constant weight, during which time several liters of water will have passed through the crucible. Another method for separating the sulphate of lime is by treatment in a weighed Gooch crucible with a mixture of nine parts of 95 per cent. alcohol and one part of glacial acetic acid. Acetates of lead, zinc, and lime being soluble in this mixture, the residue contains all the sulphate of lime and any sulphate of lead and barytes which may be present. Determine the lead and lime as in sample 4, and calculate to sulphates. Sulphate of lime should be fully hydrated in paints. To determine this, obtain loss on ignition; deduct carbonic acid and water in other constituents; the remainder should agree fairly well with the calculated water in the hydrated sulphate of lime, if it is fully hydrated. If, after washing a small portion of the sample with water, the residue shows no sulphuric acid soluble in ammonium acetate, the sulphate of lime may be obtained by determining the sulphuric acid soluble in ammonium acetate and calculating to sulphate of lime. The difficulty is in determining the sulphate of lime in the presence, or possible presence, of sulphate of lead. To illustrate the analysis of samples of white paint containing sulphate of lime and the difficulty attending thereon, we would mention a sample containing sublimed lead, white lead, carbonate of lime, and sulphate of lime. In such a sample we would determine the lead, lime, sulphuric acid, carbonic acid, loss on ignition, the portion soluble in water, and the lime or sulphuric acid in that portion, calculating to sulphate of lime. Deduct the lime in the sulphate of lime from the total lime, and calculate the remainder to carbonate of lime; deduct the carbonic acid in the carbonate of lime from the total carbonic acid, and calculate the remainder to white lead; deduct the sulphuric acid in the sulphate of lime from the total sulphuric acid, and calculate the remainder to sulphate of lead. The lead unaccounted for as sulphate or white lead is present as oxide of lead. Deduct the carbonic acid and water in the carbonate of lime and white lead from the loss on ignition, the remainder being the water of hydration of the sulphate of lime.

"*Sample* 8 contains as insoluble matter, barytes, china clay, and silica.

"After igniting and weighing the insoluble matter, carbonate of soda is added to it, and the mixture fused. The fused mass is treated with water, and the insoluble portion filtered off and washed. This insoluble portion is dissolved in dilute hydrochloric acid, and the barium present precipitated with sulphuric acid in excess. The barium sulphate is filtered out, ignited, weighed, and if this weight does not differ materially—say by 2 per cent.—from the weight of the total insoluble matter, the total insoluble matter is reported as barytes. If the difference is greater than this, add the filtrate from the barium sulphate precipitate to the water-soluble portion of fusion. Evaporate and determine the silica and the alumina in the regular way. Calculate the alumina to China clay on the arbitrary formula $2SiO_2$, Al_2O_3, $2H_2O$, and deduct the silica in it from the total silica, reporting the latter in a free state. It is to be borne in mind that china clay gives a loss of about 13 per cent. on ignition, which must be allowed for. China clay is but slightly used in white paints as compared with barytes and silica."

"*Sample* 9 contains sulphide of zinc.

"Samples of this character are usually mixtures in varying proportions of barium sulphate, sulphide of zinc, and oxide of zinc. Determine barytes as matter insoluble in nitric acid, the total zinc as in Case 1, and the zinc soluble in acetic acid, which is oxide of zinc. Calculate the zinc insoluble in acetic acid to sulphide."

"*Sample* 10 contains sulphite of lead.

"This is of rare occurrence. Sulphite of lead is insoluble in ammonium acetate, and may be filtered out and weighed as such. It is apt on exposure to the air in the moist state to become oxidized to sulphate of lead.

"There are certain arbitrary positions which the chemist must take in reporting analyses of white paints:

"*First.*—White lead is not uniformly of the composition usually given as theoretical ($2PbCO_3$), (PbH_2O_2), but in reporting we must accept this as the basis of calculating results, unless it is demonstrated that the composition of the white lead is very abnormal.

"*Second.*—In reporting oxide of lead present this should not be done except in the presence of sulphate of lead, and if white lead is present, then only where the oxide is more than 1 per cent.; otherwise calculate all the lead soluble in acetic acid to white lead.,

"*Third.*—China clay is to be calculated to the arbitrary formula given.

"In outlining the above methods we have in mind many samples that we have analyzed, and the combinations we have chosen are those we have actually found present."

General Methods for the Analysis of a Mixed Colored Paint.—Such a paint may contain any of the constituents mentioned under the various colored pigments. A qualitative examination of the paint will reveal the composition of the color present, and the constituents of such a color are then determined as specifically stated under the analysis of that particular colored pigment. The general analysis of the pigment, excepting that of the coloring portion, is carried out as stated under the analysis of a mixed white paint.

CHAPTER III.

ANALYSIS OF PAINT VEHICLES AND VARNISHES.

The can of paint is allowed to stand until the pigment has thoroughly settled, leaving the vehicle floating on top. A quantity of this separated vehicle is then removed from the can and carefully preserved in an air-tight bottle, after determining its specific gravity. A weighed portion of the vehicle may then be placed in a tared flask and attached to a Liebig condenser. Heating to 200° C. will drive off the volatile constituents. The composition of the distillate may be determined by following the methods outlined elsewhere. The residue (oil, drier and gums) may be transferred to a crucible and ignited. The residue from the ignition may then be weighed and calculated to ash. An analysis of this ash for lead, manganese and other driers, may then be proceeded with in the usual manner.

When a paint is very thick and will not settle so that the analyst cannot secure a good sample of the vehicle, a weighed portion of the paint may be placed in the thimble of a Soxhlet extractor and extraction of the vehicle made in the usual manner, with a known amount of solvent.

Some operators prefer to take a weighed quantity of the paint, as it comes from the can, say, 100 grams, and place it in a copper flask, mixing it thoroughly therein with sand. Distillation of this mass will drive over the water and volatile constituents which will separate in two distinct layers in the graduate in which they are collected. For the separation of very fine pigments, such as certain colors in oil, a Gooch crucible* is useful: Successive extractions of the pigment are decanted through a carefully prepared Gooch crucible, using a heavy bed of very fine asbestos, with a fairly strong suction.

Water.—Leo Nemzek of the North Dakota Agricultural College, who has had a wide experience in the analysis of paint,

* Warren I. Keeler, Jour. Indus. and Engineer. Chem., Vol. 2, No. 9, p. 388.

recommends the following method for the determination of water in ready-mixed paint:

Weigh out 100 grams of the paint into a 300 c.c. Erlenmeyer flask, and heat gradually after having added about 75 c.c. of toluol. The heat should not exceed 105° C. Distil over about 50 c.c. and read percentage of water. This method does not take in the water which such pigments as white lead carbonate, calcium sulphate, etc., give up at 150° C.

LINSEED OIL.

A good linseed oil will analyze within the following limits, as the results of several samples analyzed by the authors show:

Specific gravity at 15.5° C.,	.932 to .935
Acid number,	5 to 7
Saponification value,	187 to 192
Unsaponifiable matter,	.8 to 1.5%
* Iodine number,	180 to 190

The analytical methods outlined by Walker† are most excellent. Close adherence to these methods have given satisfactory results on a long series of tests conducted by the authors.

1. Preparation of Sample.

"All tests are to be made on oil which has been filtered through paper at a temperature of between 15° and 30° C. immediately before weighing, with the exception of tests No. 6, Turbidity; No. 7, Foots; No. 9, Moisture and Volatile Matter, and No. 10, Ash. The sample should be thoroughly agitated before the removal of a portion for filtration or analysis.

2. Specific Gravity.

"Determine with a pyknometer, plummet, or hydrometer at 15.5° C.

3. Viscosity.

"Use the Engler-Ubbelohde method, making the determination at 20° C.

* (May sometimes be as low as 160.)

† P. H. Walker, Some Technical Methods of Testing Miscellaneous Supplies, Bulletin No. 109, revised, Bureau of Chemistry, U. S. Dept. of Agriculture, 1910, pp. 11, 12 and 13.

4. Flash Point, Open Cup.

"Set a nickel crucible 60 mm. in diameter at the top, 40 mm. in diameter at the bottom, and 60 mm. in height in a hole in the middle of a sheet of asbestos board 200 mm. square. The bottom of the crucible should project about 25 mm. through the asbestos. Support the asbestos on a tripod and suspend a thermometer reading to 400° C. in degrees in the center of the crucible, so that the lower end of the thermometer is 10 mm. from the bottom of the crucible. Then pour in the oil until its level is 15 mm. below the top of the crucible. Place a Bunsen burner below the crucible and regulate the size of flame so that the thermometer rises 9° a minute. As a test flame use an ordinary blowpipe attached to a gas tube. The flame should be about 6 mm. long. Begin testing when the temperature of the oil reaches 220° C., and test for every rise of 3°. In applying the test move the flame slowly across the entire width of the crucible immediately in front of the thermometer and 10 mm. above the surface of the oil. The flask point is the lowest temperature at which the vapors above the oil flash and then go out.

5. Fire Point.

"After noting the temperature at which the oil flashes continue the heating until the vapors catch fire and burn over the surface of the oil. The temperature at which this takes place is the fire point. In determining the flash point note the behavior of the oil. It should not foam or crack on heating Foaming and cracking are frequently caused by the presence of water.

6. Turbidity.

"Note whether the oil is perfectly clear or not.

7. Foots.

"Let a liter of the oil stand in a clear glass bottle for eight days, and then note the amount of sediment formed. The highest grades of oil show no turbidity or foots by this test. The claim is made that sometimes what would be called foots by the above method is due to the freezing out of fats of rather high melting point. When a sufficient amount of the sample is available, heat one portion to 100° C. and set it aside for the determination of foots, together with a sample just as it is received. Note also the odor of the warm oil, rubbing it on the hands; a small amount of fish oil may be detected in this way.

8. Break.

"Heat 50 c.c. of the oil in a beaker to 300° C. Note whether the oil remains unchanged or "breaks;" that is, shows clots of a jelly-like consistency.

9. Moisture and Volatile Matter.

"Heat about 5 grams of oil in an oven at 105° for forty-five minutes; the loss in weight is considered as moisture. This determination is of course not exact, as there is some oxidation. When a more accurate determination is desired, perform the whole operation in an atmosphere of hydrogen.

10. Ash.

"Burn about 20 grams of oil in a porcelain dish and conduct the ashing at as low a temperature as possible. The best oil should contain only a trace of ash. An amount as large as 0.2 per cent. would indicate an adulterated or boiled oil. Examine the ash for lead, manganese, and calcium.

11. Drying on Glass.

"Coat glass plates 3 by 4 inches with the oils to be examined, expose to air and light, and note when the film ceases to be tacky. A good oil should dry to an elastic coherent film in three days. Varying conditions of light, temperature, and moisture have such an influence on drying tests that for comparison of one linseed oil with others all samples must be run at the same time.

12. Drying on Lead Monoxide.

"Livache's test calls for precipitated lead, but litharge gives equally good results. Spread about 5 grams of litharge over the flat bottom of an aluminum dish 2.5 inches in diameter and 5/8 inch high; weigh the dish and the litharge; distribute as evenly as possible over the litharge 0.5 to 0.7 gram of the oil, weigh exactly, expose to the air and light for ninety-six hours, weigh again, and calculate the gain in weight to percentage based on the original weight of the oil used.

13. Acid Number.

"Weigh 10 grams of oil in a 200 c.c. Erlenmeyer flask, add 50 c.c. of neutral alcohol, connect with a reflux air condenser, and heat on a steam bath for half an hour. Remove from the bath, cool, add phenolphthalein, and titrate the free acid with fifth-normal sodium hydroxide. Calculate as the acid number (milligrams of potassium hydroxide to 1 gram of oil). The acid number varies with the age of the oil, and should be less than 8, though when the oil is refined with sulphuric acid it may show a higher acid number. Test for sulphuric acid.

14. Saponification Number.

"Weigh from 2 to 3 grams of oil in a 200 c.c. Erlenmeyer flask, add 30 c.c. of a half-normal alcoholic solution of potassium hydroxide, connect with a reflux air condenser, heat on a steam bath for an hour, then titrate with half-normal sulphuric acid, using phenolphthalein as indicator. Always run two blanks with the alcoholic potash. From the difference between the number of cubic centimeters of acid required by the blanks and the determinations, calculate the saponification number (milligrams of potassium hydroxide to 1 gram of oil). The saponification number should be about 190.

15. Unsaponifiable Matter.

"As the saponification varies somewhat in pure oil, it is sometimes advisable to make a direct determination of unsaponifiable matter. Saponify from 5 to 10 grams of oil with alcoholic potassium hydroxide (200 c.c. of a half-normal solution) for an hour on a steam bath, using a reflux condenser. Then remove the condenser and evaporate the alcohol as completely as possible; dissolve the soap in 75 c.c. of water, transfer to a separatory funnel, cool, shake out with two portions of 50 c.c. each of gasoline distilled between 35° and 50° C., wash the gasoline twice with water, evaporate the gasoline, and weigh the unsaponifiable matter, which in raw linseed oil should be below 1.5 per cent.; in boiled oil it is somewhat higher, but should be below 2.5 per cent.

16. Iodine Number.

"Weigh in a small glass capsule from 0.2 to 0.25 gram of oil, transfer to a 350 c.c. bottle having a well-ground stopper;

dissolve the oil in 10 c.c. of chloroform and add 30 c.c. of Hanus solution; let it stand with occasional shaking for one hour, add 20 c.c. of a 10 per cent. solution of potassium iodide and 150 c.c. of water, and titrate with standard sodium thiosulphate, using starch as indicator. Blanks must be run each time. From the difference between the amounts of sodium thiosulphate required by the blanks and the determination, calculate the iodine number (centigrams of iodine to 1 gram of oil). The iodine number of raw linseed oil varies from 175 to 193, though Gill states that a pure raw oil may give a value as low as 160. Boiled oil may be very much lower.

"Make the Hanus solution by dissolving 13.2 grams of iodine in 1,000 c.c. of glacial acetic acid which will not reduce chromic acid, and adding 3 c.c. of bromine.

17. Rosin or Rosin Oil (Liebermann-Storch Test).

"To 20 grams of oil add 50 c.c. of alcohol, heat on a steam bath for fifteen minutes, cool, decant the alcohol, evaporate to dryness, add 5 c.c. of acetic anhydride, warm, cool, draw off the acetic anhydride, and add a drop of sulphuric acid, 1.53 specific gravity. Rosin or rosin oil gives a fugitive violet color."

The hexabromide test is sometimes of value. The method used by Committee E in their work on linseed oil follows:*

"**Hexabromide Test.**—On oil, determining the melting point of the bromide compounds. Method to be followed: The determination should be made in glass-stoppered, weighing bottles, about 6 inches high and 1 inch in diameter, with flat bottom, and weighing about 30 grams each. These bottles should be carefully dried and weighed. Weigh into one of these bottles 0.3 gram of oil to be tested; add 25 c.c. of absolute ether; cool to near 0° C.; add bromine, drop by drop, until a considerable excess is shown by the color of the solution. Stir constantly during this addition, and add bromine very slowly to avoid heating. Place tube in ice water for thirty minutes; then in centrifuge, whirling for two minutes at speed of 1,200 revolutions per minute. This throws the brominated oil to the bottom of the tube, from which the supernatant liquid can be easily and quickly decanted. Add 10 c.c. of cold ether; stir precipitate with glass rod; allow to stand in ice water until

* Proc. Amer. Soc. for Testing Materials, Vol. IX, p. 152.

thoroughly cold. Whirl in centrifuge again, and decant supernatant liquor. Another washing in the same manner will remove the excess of bromine and oil. Allow the tube and residue to stand for a short time, until the ether has evaporated; dry in water bath for thirty minutes, and weigh (Tolman's method)"

Acetyl Value.

The determination of acetyl value is based on the principle that hydroxy acids on being heated with acetic anhydride exchange the hydrogen atom of their hydroxy group or groups for the radicle of acetic acid. The procedure is as follows:

Boil the oil with an equal volume of acetic anhydride for two hours in a round-bottomed flask attached to an inverted condenser, transfer to a large beaker, mix with several hundred cubic centimeters of water and boil for half an hour.

A slow current of CO_2 should be passed into the liquid through a finely drawn-out tube reaching nearly to the bottom of the beaker; this is done to prevent bumping. The mixture is allowed to separate into two layers, the water is siphoned off and the oily layer again boiled out until the last trace of acetic acid is removed, which can be ascertained by testing with litmus paper. The acetylated product is freed from water and finally filtered through filter-paper in a drying oven.

*"This operation may be carried out quantitatively, and in that case the washing is best done on a weighed filter. On weighing the acetylated oil or fat, an increase of weight would prove that assimilation of acetyl groups has taken place. This method may be found useful to ascertain preliminarily whether a notable amount of hydroxylated acids is present in the sample under examination.

"Two or 4 grams of the acetylated substance are saponified by means of alcoholic potash solution, as in the determination of the saponification value. If the 'distillation process' be adopted it is not necessary to work with an accurately measured quantity of standardized alcoholic potash. In case the 'filtration process' be used, the alcoholic potash must be measured exactly. (It is, however, advisable to employ in either case a

*Commercial Organic Analysis, by Alfred H. Allen, pp. 66 and 67. P. Blakiston's Son & Co., Philadelphia, 1905.

known volume of standard alkali, as one is then enabled to determine the saponification value of the acetylated oil or fat.) Next the alcohol is evaporated and the soap dissolved in water. From this stage the determination is carried out either by the (*a*) 'distillation process' or (*b*) 'filtration process.'

"(*a*) *Distillation Process.*—Add dilute sulphuric acid (1:10), more than sufficient to saturate the potash, and distil as usual in Reichert's distillation process. Since several 100 c.c. must be distilled off, either a current of steam is blown through the suspended fatty acids or water is run into the distilling flask, from time to time, through a stoppered funnel fixed in the cork, or any other convenient device is adopted. It will be found quite sufficient to distil over 500 to 700 c.c., as the last 100 c.c. contain practically no acid. Filter the distillates to remove any insoluble acids carried over by the steam, and titrate the filtrates with decinormal potash, phenolphthalein being the indicator. Multiply the number of cubic centimeters by 5.61 and divide the product by the weight of substance taken. This gives the acetyl value.

"(*b*) *Filtration Process.*—Add to the soap solution a quantity of standardized sulphuric acid exactly corresponding to the amount of alcoholic potash employed and warm gently, when the fatty acids will readily collect on the top as an oily layer. (If the saponification value has been determined it is, of course, necessary to take into account the volume of acid used for titrating back the excess of potash.) Filter off the liberated fatty acids, wash with boiling water until the washings are no longer acid, and titrate the filtrate with decinormal potash, using phenolphthalein as indicator. The acetyl value is calculated in the manner shown above.

"Both methods give identical results; the latter will be found shorter.

"The acetyl value indicates the number of milligrams of KOH required for the neutralization of the acetic acid obtained on saponifying 1 gram of the acetylated fat or wax."

Maumene Test.

While this test is not strictly a quantitative one, the indications afforded by it are of considerable value. It depends on the heat developed by the mixing of the oil with strong sulphuric acid and is carried out as follows:

A beaker of about 150 c.c. capacity and from 7 1/2 to 9 cm. deep is carefully placed into a larger vessel and surrounded with dry felt or cotton waste.

Fifty grams of the oil are put into the beaker and 10 c.c. of concentrated sulphuric acid are gradually introduced into the oil from a burette and the mixture stirred until no further increase in temperature is recorded by a thermometer immersed in it.

The highest point at which the thermometer remains constant for any appreciable time is observed, and the difference between this and the initial temperature is the rise of temperature.

In performing this test, it is highly important that the oil and acid be originally of the same temperature and that the strength of the acid should be the same as far as possible.

As the rise in temperature varies with the strength of the acid, to secure uniformity the results should be expressed by dividing the rise of temperature with the oil by the rise of temperature with water and multiplying by 100. This is called the specific temperature reaction.

The rise of temperature with water is determined in the same manner as with oil, using the same vessel.

Raw linseed oil gives a Maumené number about 20° F. in excess of that given by some boiled linseed oils.

Polarimetric Test for Rosin Oil with Linseed Oil.

*"The investigations of Bishop and Peters on the opticity of a number of oils show that, with the exception of caster oil, croton oil and rosin oil, the only dextrorotations are produced by sesame (high) and olive oil (feeble), all the others, including linseed oil, being either optically inactive or having a light levo-rotatory power.

"(1) *The Polarimetric Examination of Linseed Oil Sophisticated with Refined Rosin Oil* (R. R. O.).—According to Aignan such a mixture rotates the plane of polarization to the right by an angle perceptibly proportional to the quantity of rosin oil which it contains. If the rotation observed with a 200 mm. tube be represented by a D and the weight of the rosin oil in 100

*The Manufacture of Varnishes, and Kindred Industries, Livache and McIntosh, Vol. I, D. Van Nostrand Company, New York, 1904.

parts by weight of the mixture by h, we get in the case of a mixture of linseed oil with

1. Refined rosin oil, [a] D + = 14/15 h.
2. Choice white rosin oil, [a] D + = 17/15 h.
3. Rectified rosin oil, [a] D + = 21/15 h.

"The first mixture is the most common. In actual practice, therefore, all that has to be done is to measure a D by the polarimeter, and to estimate h as refined rosin oil, according to the formula h = [a]D 15/14. The oils in question being dark in color, it is better to work in a 100 mm. tube and to calculate

$$h = [a]D = 15/7.$$

"(2) *Estimation of Rosin Oil in Paint by the Polarimeter.*—(A) A certain amount of the paint is frequently stirred and shaken up with ether and allowed to settle. The ether containing the oil in solution floats to the surface and the polarization tube is filled with the ethereal solution. If no optical deviation be produced, there is no rosin oil in the paint tested. On the other hand, if [a]D be the rotation toward the right with a 200 mm. tube, according to Aignan's researches on the rotatory power of an ethereal solution of linseed oil containing rosin oil, the proportion of rosin oil may be calculated by the formula

$$h = \frac{a[D]}{43^2}.$$

"(B) A known weight, p^1, of the ethereal solution is run into a flask and heated on the water-bath at 100° C. (212° F.) so as to drive off the ether; the oil which boils only at 300° C. (572° F.) is left in the flask. Let its weight be represented by p^2. The proportion $\frac{p^1}{p^2} 100 = h^1$ per cent. of oil (linseed oil and rosin oil) contained in the ethereal solution examined by the polarimeter. If $h^1 = h$, it may be taken for granted that the paint contained linseed oil free from rosin oil. Generally, h^1 is greater than h, then $\frac{h}{h^1} 100$ will give the percentage of rosin oil contained in the linseed oil which was used to make the paint."

The detection of other vegetable or animal oil, admixed with linseed oil, is not always an easy task when they are present in small percentage. Petroleum oil and rosin oil are the most common adulterants. The detection and estimation of the former is given on page 52.

The presence of rosin oil as an adulterant may be detected by shaking the oil with an equal quantity of acetic anhydride. Acetic anhydride is removed from the oil and mixed with a few drops of concentrated sulphuric acid. Production of a violet color indicates the presence of a rosin oil. Its specific gravity is very high and saponification number very low.

Lewkowitsch* shows the determination of rosin acids in admixture with fatty acids, as follows:

"Twitchell's method is based on the property aliphatic acids possess of being converted into their ethylic esters when acted upon by hydrochloric acid gas in their alcoholic solution, whereas colophony practically undergoes little change under the same treatment, abietic acid separating from the solution. The analysis is carried out as follows:

"Two to three grams of the mixed fatty and rosin acids are weighed off accurately, dissolved in a flask in ten times their volume of *absolute* alcohol (90 per cent. alcohol must not be used, as the conversion of fatty acids into esters is not complete in that case), and a current of dry hydrochloric acid gas passed through, the flask being cooled by immersion in cold water. The gas is rapidly absorbed at first, and after about forty-five minutes, when unabsorbed gas is noticed to escape, the operation is finished. To ensure complete esterification the flask is allowed to stand for an hour, during which time the ethylic esters and the rosin acids separate on the top as an oily layer. The contents of the flask are then diluted with five times their volume of water, and boiled until the aqueous solution has become clear. From this stage the analysis may be carried out either (a) volumetrically or (b) gravimetrically.

"(a) *The Volumetric Analysis*—The contents of the flask are transferred to a separating funnel, and the flask rinsed out several times with ether. After vigorous shaking the acid layer is run off, and the remaining ethereal solution, containing the ethylic esters and the rosin acids, washed with water until the last trace of hydrochloric acid is removed. Fifty c.c. of alcohol are then added, and the solution titrated with standard

* Chemical Technology and Analysis of Oils, Fats, and Waxes, by Lewkowitsch, Vol. I, p. 394. The Macmillan Co., New York, 1904.

caustic potash or soda, using phenolphthalein as an indicator. The rosin acids combine at once with the alkali, whereas the ethylic esters remain practically unaltered. Adopting as the combining equivalent for rosin 346, the number of c.c. of normal alkali used multiplied by 0.346 will give the amount of rosin in the sample.

"(b) *The Gravimetric Method.*—The contents of the flask are mixed with a little petroleum ether, boiling below 80° C., and transferred to a separating funnel, the flask being washed out with the same solvent. The petroleum ether layer should measure about 50 c.c. After shaking, the acid solution is run off, and the petroleum ether layer washed once with water, and then treated in the funnel with a solution of 0.5 grm. potassium hydroxide and 5 c.c. of alcohol in 50 c.c. of water. The ethylic esters dissolved in the petroleum ether will then be found to float on the top, the rosin acids having been extracted by the dilute alkaline solution to form rosin soap. The soap solution is then run off, decomposed with hydrochloric acid, and the separated rosin acids collected as such, or preferably dissolved in ether and isolated after evaporating the ether. The residue, dried and weighed, gives the amount of rosin in the sample."

The detection of olive oil, palm oil, cocoanut oil, cotton-seed oil, fish oil, and other vegetable and animal oils may be made by following the methods in "Oil Analysis," by Gill (Lippincott Co.). The low iodine values of the above-named oils as compared to that of linseed oil is generally sufficient evidence of their presence.

SOYA BEAN OIL.

Soya bean oil in its chemical constants runs so close to linseed oil that it is very hard to detect. The same methods can be used, however, as for the analysis of linseed oil. When working upon mixtures of the two the following table of the authors will probably be valuable in their identification.

Chemical Characteristics Of Soya Bean Oil.

Sample No.	Specific gravity	Acid No.	Saponification No.	Iodine No.	Per cent. of foots.
1	0.9233	1.87	188.4	127.8	3.81
2	0.924	1.92	188.3	127.2	
3	0.9231	1.90	187.8	131.7	
4	0.9233	1.91	188.4	129.8	
5	....	....	....	130.0	
6	....	....	....	132.6	
7	....	....	....	136.0	
Average ..	0.9234	1.90	188.2	130.7	

It is evident that the iodine value of soya bean oil is the only chemical characteristic that markedly differentiates it from linseed oil. Therefore, in the detection of soya bean oil and its estimation, the iodine values of several samples of mixed oils are given as being of interest in this connection:

Iodine Values Of Linseed Oil And Mixed Oils.

Sample No.	Straight linseed	25% soya. 75% linseed	50% soya. 50% linseed	75% soya. 25% linseed
-	190.3	175.2	160.7	140.4
2	189.5	175.9	161.7	140.8
3	188.0	175.4	160.3	139.9
Average	189.3	175.5	160.9	140.4

The authors have found that treatment of a few drops of soya bean oil, or oil containing any considerable percentage of soya bean oil, with one drop of concentrated sulphuric acid will produce a distinct fluorescent yellowish-green color. This color is entirely different from that produced with pure linseed oil, which is of a brownish-red and of a begonia-shaped pattern. This test is best conducted on the lid of a porcelain crucible. Subsequent examination under the microscope is of value in confirming the test. The comparatively slow drying of soya bean oil will often indicate its presence.

CHINESE WOOD OIL.

Investigations, extending over several years, conducted by Kreikenbaum,* determined that Chinese wood oil as it comes to the paint trade is fairly uniform in its constants. Kreikenbaum's work also determined that the Hanus method for the determination of the iodine number, although applicable in the case of linseed oil, could not be used when working on Chinese wood oil, as it gave abnormally high results. The average constants of a large number of commercial samples determined by Kreikenbaum follow:

Specific gravity	.941 to .943
Free acid,	4.4
Saponification number,	190.9
Hubl iodine number,	169 to 171

Working with the Hübl method a six-hour absorption is sufficient to get good accurate results when determining the iodine number of this oil.

SPIRITS OF TURPENTINE, PETROLEUM AND LIGHT OILS.

For a quick test to determine whether a turpentine is pure, the chemist may mix in a test-tube 10 c.c. of the material under examination and 10 c.c. of aniline oil. If the turpentine is pure, the two materials will mix without turbidity. If petroleum products are present, they will be indicated by a cloudiness and quick separation in a distinct layer from the turpentine and aniline oil.

The following methods have been given by Walker† for the analysis of the volatile solvents used to a great extent in the manufacture of paints:

* Adolph Kreikenbaum. Constants of Chinese Wood Oil, Vol. II, No. 5, Jour. Indus. and Engineer. Chem.

† P. H. Walker. Some Technical Methods of Testing Miscellaneous Supplies, Bull. 109, revised, Bureau of Chemistry, U. S. Dept. of Agriculture, 1910, pp. 13, 14, and 15.

SPIRITS OF TURPENTINE.*

1. Color.

"The best quality of spirits of turpentine should be water-white.

2. Specific Gravity.

"Determine the specific gravity with a pyknometer, plummet, or hydrometer at 15.5° C. Pure gum turpentine should have a density between 0.862 and 0.875. Wood turpentine may, however, range from 0.860 to 0.910 or even higher.

3. Distillation.

"Connect a distilling flask of 150 c.c. capacity with a condenser having a thermometer. Introduce 100 c.c. of turpentine and heat with a Bunsen burner. The initial boiling point should be about 156° C., and 95 per cent. should distil over between 153.5° and 165.5° C.

4. Residue on Evaporation.

"Evaporate 10 grams on the steam-bath; the residue should be less than 2 per cent.

5. Refractive Index.

"Determine with a Zeiss direct reading refractometer at 20° C. The index of refraction for gum turpentine should be from 1.4690 to 1.4740; for wood turpentine, 1.4685 to 1.5150.

6. Action of Sulphuric Acid (Polymerization).

"Measure 6 c.c. of turpentine in a stoppered, thin-walled tube graduated to 0.1 c.c. (carbon tubes). Place the tube in cold water and pour in slowly a mixture of four parts of strong sulphuric acid and one part of fuming sulphuric acid. Add the acid slowly, and avoid an excessive rise in temperature. Shake the tube so as to mix the turpentine and the acid, add finally about 20 c.c. of the acid, stopper the tube, mix thoroughly, cool, allow to stand thirty minutes, and note the volume of unpolymerized oil that collects on top of the acid layer. Then

* If wood turpentine has been carefully refined, it will comply with all the tests given for spirits of turpentine, but it can almost invariably be distinguished from the latter by its characteristic odor.

let stand for eighteen hours and again note the volume. A pure turpentine should show less than 0.3 c.c. unpolymerized at the end of thirty minutes, and less than 0.5 c.c. after eighteen hours.

"This method will indicate gross adulteration, but will not detect admixtures of very small amounts of mineral oil. Donk has perfected a method which determines the presence of as little as 1 per cent. of mineral oil in turpentine. This method is as follows:

"Sulphuric acid of thirty-eight times the normal strength (101.5 per cent.) is prepared by mixing very strong sulphuric acid with fuming sulphuric acid. It must be determined by titration that this reagent is of the exact strength required, for with 37.5 times normal acid (100 per cent.) the turpentine is not completely destroyed, and with acid stronger than 101.5 per cent. the amount of mineral oil dissolved becomes excessive.

"Place about 25 c.c. of the special sulphuric acid in a flask having a narrow graduated neck (a Babcock bottle does very well), cool in ice-water, add 5 c.c. of the turpentine to be tested and cool the flask again, shaking it carefully and avoiding any excessive rise in temperature by frequent cooling. The flask should never be too hot to hold in the palm of the hand. Then place it in a bath of cold water and heat the bath at such a rate that in about five minutes the temperature will be 65° C. During the heating shake the bottle about every fifty seconds, finally shaking very thoroughly so as to insure the contact of every particle of the sample with the acid. Cool to room temperature and add ordinary strong sulphuric acid in sufficient amount to bring the unpolymerized liquid up in the graduated neck. Let stand overnight or whirl in a centrifuge and read the volume on the neck.

"Pure turpentine should leave a residue of not over 0.04 c.c., which is not limpid and which has a refractive index of not less than 1.500.

"If the unpolymerized residue is 0.04 c.c. or less, mineral oil may be assumed to be absent. If the residue is greater, calculate from it the percentage of mineral oil present. This will be, of course, only approximate, for there is some residue from pure turpentine and some mineral oil is dissolved by the acid; but for all practical purposes it may be assumed that the errors balance one another, and hence it is not advisable to apply any correction.

7. Spot Test.

"Place a drop on filter-paper and allow it to dry at room temperature; it should leave no stain.

8. Flash Point.

"Support a crucible, such as is used in determining the flash point of linseed oil, in a vessel of water at 15° to 20° C.; the water should cover about two-thirds of the crucible. Fill the crucible to within about 2 cm. of the top with turpentine, insert a thermometer, and heat the water bath slowly, 1° per minute. Begin at 37° and test for the flash at each rise of 0.5°. The turpentine should not flash under 40.5° C.

Wood Turpentine.

The steam method* for the distillation of wood turpentine, as worked up by Geer, Bristol, Hawley, and others of the Forest Products Laboratory of the United States, separates the various high and low boiling-point fractions, all of which have different characteristics.

BENZINE AND LIGHT PETROLEUM OILS.

"The term benzine is used for a number of light petroleum oils. In the painting trade it generally refers to a product of about 62° Baumé (0.7292 sp. gr.). The petroleum benzine of the U. S. Pharmacopœia is a lighter oil, being a light gasoline.

1. Specific Gravity.

"Determine with a spindle, pyknometer, or plummet at 15.5° C. The determination can be made at room temperature and corrected to 15.5° C.

2. Sulphur (Sodium Nitroprusside Test).

"To 100 c.c. of the sample in a flask add about 1 gram of bright metallic sodium, connect with a reflux condenser, and boil for one hour. Cool, add water drop by drop until the metal

* The Analysis of Turpentine by Fractional Distillation with Steam, by Wm. C. Geer, U. S. Dept. of Agriculture, Forest Service Circular 152.

is dissolved, separate the aqueous liquid, and test with a drop of sodium nitroprusside solution. A fine violet-blue coloration indicates sulphur.

3. Sulphur Compounds and Pyrogenous Products (U. S. P. Test).

"To 100 c.c. of the sample add 25 c.c. of a solution of 10 per cent. anhydrous ammonia in 95 per cent. alcohol (spirit of ammonia U. S. P.), add 1 c.c. of silver nitrate solution. Boil gently for five minutes. A brown coloration indicates sulphur compounds or pyrogenous products.

4. Residue on Evaporation.

"Place 25 c.c. in a 100 c.c. platinum dish, heat on steam-bath for thirty minutes, and weigh residue. No residue should be left by this test.

5. Fractional Distillation.

"Light petroleum oils are usually tested only for specific gravity; but as light and heavy distillates may be mixed, the specifications would be improved by requiring that a certain fraction should distil between specified temperatures. To make this determination distil 100 c.c. in a round-bottom flask 6.5 cm. in diameter; the neck should be 1.6 cm. in diameter and 15 cm. long, with a side tube set in the middle of the neck at an angle of 75°. The surface of the liquid should be 9 cm. below the side tube, and the bulb of the thermometer just below the side tube.

6. Benzol.

"Mix the sample with 8 volumes of strong sulphuric acid and 2 volumes of nitric acid; heat gently for ten minutes, allow to cool, and note odor. The odor of nitrobenzol indicates benzol.

7. Color and Odor.

"Note color of sample and odor both in bulk and after rubbing on hands."

ANALYSIS OF VARNISH AND JAPAN.

An examination of a varnish should be largely of a physical nature, the chemist determining its appearance, odor, body, clarity, and properties of the dried film. The specific gravity, flash point, viscosity, acid number, ash, and rosin test, as well as the percentage of volatile oils, are determined in the usual manner, as outlined for linseed oil on previous pages. The percentage of fixed oils and gums is determined by weighing that portion left in the flask after distillation of the volatile constituents. Owing to the chemical combinations and changes which are effected when various gums are heated together in the presence of oils, it is almost impossible for the most expert analyst to determine the exact make-up of a varnish. The analyst, however, may determine the total percentage of gums and the percentage total of oils with a fair degree of accuracy, on an original sample by precipitation of the insoluble gums with gasoline, determining the soluble gums in the benzine by extraction with chloroform after evaporation of the gasoline and oxidation of the oil. To make an exact determination of a varnish formula, however, as it was submitted to the varnish maker, is almost impossible.

Oils, Gums and Volatile.—In ordinary varnish which generally contains oil, gums and volatile solvents like benzine and terpentine, weigh off 10 grams into a tall thin beaker of 400 c.c. capacity. Next add a large amount of ice cold petroleum ether. Cover the beaker and allow to stand for several hours when the gums will be found separated at the bottom of the beaker. Repeat the extraction with cold petroleum ether at least three times, pouring the several decanted portions into a large bottle.

After finishing the extraction, add 100 c.c. of ice cold water to the petroleum ether in the bottle and shake thoroughly, causing a small amount of gum which usually dissolves in the ether to reprecipitate. Filter on a tared filter, previously moistened with ice water, and also transfer to this filter the gum contained in the beaker, using a stirring rod and some petroleum ether to loosen it from the glass, washing finally with a small quantity of ice water. Dry at 100 C. and weigh as gums.

If the oil is also to be determined, the petroleum ether and water can be separated in a separatory funnel and the ether then evaporated or distilled off, leaving the oil.

The sum of the oil and gums subtracted from 100 gives the percentage of volatile thinners—benzine, turpentine, etc.

Distillation of the volatile solvents from a separate portion of the varnish may be used to determine the percentage of turpentine by the polymerization method previously outlined in this chapter.

In spirit varnishes, like shellac, to determine the gums it is only necessary to weigh off 10 grams into a tared porcelain dish and evaporate off the solvent on the water bath.

McIlhiney's method for the analysis of oil varnishes has been used by the authors with success. This method may also be used in the analysis of japan driers.

McIlhiney's Method.—Of the materials used in construction few if any present more difficult problems in their testing and analysis than oil varnishes. Any rational system of testing varnishes to determine their suitability for a given use and their resistance to the destructive effects of exposure to the elements, will take account among other tests of a chemical analysis to determine the ingredients of the varnish and the proportions in which they are combined. It is unfortunate that there is not at the present time any method of analysis which will determine with any reasonable degree of accuracy, the proportions in which the oil, hard gum, and common rosin have been combined to form the nonvolatile base from which the varnish is produced by dilution with turpentine or other volatile oil. The proportion of volatile oil in the varnish may be determined by distilling the solvent off with steam at a temperature a little above the boiling point of water and then separating the volatile oil from the aqueous part of the distillate and weighing or measuring it. Its further examination need not be entered upon here since methods of analysis of such volatile oils as are likely to be used as thinners for varnish are now generally known and are described at length in such standard works as Allen's Commercial Organic Analysis, and the books on paints and varnishes by Toch, Sabin, and Holley and Ladd.

The separation of the hard gum from the oil and the common rosin is the problem which is difficult; the hard gum and the oil do not unite at all, practically, until the hard gum has been melted and from 15 to 25 per cent. of its weight driven off as vapor, the amount so lost depending upon the character of the gum. After this melting the linseed oil may be added if it has been previously heated enough to prevent it from chilling the melted gum. This mixture of oil and gum is usually at this stage heated for a shorter or longer time to complete the combination of the ingredients. The union of oil and hard gum which has been effected by this means cannot be broken up by any solvent, or perhaps it would be more accurate to say that after the combination between hard gum and oil has been successfully made and the mixture thinned with turpentine and stored for a few months, no solvent can be depended upon regularly to effect a separation of the two by its selective solvent action.

The process which is here described depends upon the fact that although the union between oil and hard gum is too intimate to be broken up by the selective solvent action of any solvent acting directly upon the original mixture, the combination may be broken up and the oil and gum brought back to more nearly their original condition before they were melted together, by submitting the mixture to the action of caustic potash in alcoholic solution and subsequently acidifying the solution of potash salts so formed. By this means there is obtained from hard gum varnishes a quantity of gum insoluble in petrolic ether very closely approximating the amount of hard gum actually existing in the varnishes, while the linseed oil is represented by its fatty acids which are readily soluble in this solvent unless they have been oxidized, in which case some of the fatty acids of the linseed oil will accompany the insoluble hard gum.

In carrying out the method an opportunity is given to determine not only the weight of the oil and of gum but also the Koettstorfer figure and the percentage of glycerine in the mixture. All these data taken together give a basis for corroborating the main figures.

The process is carried out by weighing into an Erlenmeyer flask 2 to 10 grams of the varnish, adding a considerable excess of approximately half-normal solution of caustic soda or caustic potash in very strong or absolute alcohol, distilling off the major portion of the solvent and redissolving in neutral absolute alcohol. The solution is then titrated with a solution of pure acetic acid in absolute alcohol, approximately half-normal strength, to determine the amount of the excess of alkali present. From this the Koettstorfer figure is determined as the exact strengths of the acid and alkali solutions have been ascertained independently by comparison with known standards. A further quantity of the standard solution of acid in alcohol is added so as to exactly neutralize the total amount of alkali originally added. By this means the acid bodies liberated from their combinations with alkali are obtained in solution in strong alcohol. To this solution there is now added a sufficient quantity of petrolic ether to dissolve the oil acids and this petrolic ether being miscible with the strong alcohol forms with it a homogeneous liquid. Water is now added to the mixture in such amount as to so dilute the alcohol contained that it is no longer a solvent for fatty or resin acids; this addition of water causes the petrolic ether which was mixed with the alcoholic liquid to separate carrying with it the fatty acids. The rosin goes with the fatty acids while the hard gum being insoluble in either the petrolic ether or in the very dilute alcohol separates in the solid state. The aqueous and ethereal layers are now separated in a separating funnel and each is washed, the watery layer with petrolic ether and the petrolic ether layer with water. The petrolic ether layer is now transferred to a weighed flask, the solvent distilled off, and the residue of fatty acids and common rosin weighed. This latter is then examined further by Twitchell's method to determine the amount of rosin which it contains or it may be examined qualitatively in a number of ways to establish its identity.

The aqueous layer is freed from the suspended hard gum which it contains by filtering, and from any further quantity of gum which the weak alcohol may have retained in solution by evaporating off the alcohol and again filtering. The remaining aqueous liquid contains the glycerin and this is determined by the Hehner method with potassium bichromate—the method ordinarily used for examining spent soap lyes.

The hard gum is, according to this plan, precipitated in such a way that it adheres to the sides of the glass vessel in which the alcohol and petrolic ether mixture is diluted with water; the easiest method to weigh it is, therefore, to carry on the operation of dilution in a weighed glass vessel and then to dry and weigh the hard gum in this vessel. It frequently happens that some of the hard gum cannot be conveniently retained in this vessel but that it must be filtered out on a weighed filter and the weight so found added to that of the main portion.

If the varnish contains nonvolatile petroleum or other

unsaponifiable matter it will naturally be included in the fatty and resin acids, and it would be necessary to saponify the latter and extract the unsaponifiable matter from them while in the alkaline state; this operation is so familiar to chemists that it is mentioned here only to call attention to the necessity for it in some cases.

It would naturally be expected that on account of the well known insolubility of the oxidized fatty acids in petrolic ether, some of the acids of the linseed oil which had been polymerized by heat during the cooking of the varnish, or which had been oxidized during the blowing process to which some linseed oil is subjected before making it up into varnish, would fail to dissolve and would be counted in with the hard gum instead of with the linseed oil. It appears as a matter of fact that this source of error is of slight importance in the case of oil thickened by heat but that the blowing process gives an oil which is not completely accounted for by the soluble fatty acids recovered. This difficulty may be largely overcome by taking advantage of the greater solubility of the oxidized fatty acids in alcohol as compared with the hard gum; the freshly precipitated gum contaminated with oxidized fatty acids is treated with a moderate quantity of cold alcohol of about 85 per cent. and allowed to digest for some time. The soluble matter so extracted is then recovered separately by evaporating off the alcohol.

The great variety of hard gums in use and in the methods of making them up into varnish make the problem one of great complexity. It is not to be expected that any one method of analysis or any single set of directions for carrying on the operation of making the analysis would be generally applicable, and it is not the intention in this paper to give such detailed instructions. The method described has, however, been found to give, upon samples of known composition made up under conditions which imitate closely the conditions of practice in an ordinary varnish factory, results that were accurate to within reasonable limits.

Rosin when present is usually combined with lime in the proportion of about one part of lime to twenty parts of rosin. An examination of the mineral constituents of the varnish is therefore of some value; the extraction of the mineral bases may be effected by treating a quantity of the varnish somewhat thinned with benzine, with strong hydrochloric acid, and examining the aqueous liquid.

The amount of fatty acids obtained represents about 92.5 per cent. of the linseed oil. The identification of these fatty acids as belonging to linseed oil or to china wood oil may be satisfactorily accomplished in some cases, but there are undoubtedly many varnishes in which the analyst will be unable to identify and determine the oils. The identification of the hard gums after separation from the other constituents of the varnish is a matter for which no rule can be given. The odor and physical characteristics of the recovered gum are quite as important as the known chemical tests of which the acidity and the Koettstorfer figure are among the most important. The chemistry of these gums is as yet almost unknown, but in the near future it is likely that our knowledge both of the nature of these hard gums and of methods for separating them from the other ingredients of the varnishes of which they form a part, will be very greatly increased.

For the analysis of gums, the analyst is referred to Vols. I and II of Lewkowitsch, and to Allen's Commercial Organic Chemistry. The following method of McIlhiney's for the analysis of shellac, being of such great value, is presented at this place:

THE ANALYSIS OF SHELLAC.

McIlhiney's Method.*—In the last few years the analytical examination of shellac has become much more common because the users of shellac and the dealers in it have become better informed as to the extent to which adulteration has been practised in the past, and particularly because more accurate and reliable methods of analysis have been devised for examining shellac. The most important adulterant of shellac, in fact almost the only adulterant, is common rosin or colophony, and it is to the detection and estimation of this adulterant that most of the analytical methods have been directed. The price of commercial shellac is determined by its color, freedom from dirt, etc., as well as by its content of colophony, but the chemist is not usually called upon to determine the commercial grade

* Presented before the International Congress of Applied Chemistry.

of shellac in other respects than its purity or freedom from adulteration.

The methods which have been used with greater or less success to determine the amount of rosin in shellac depend first upon the different behavior of shellac and of colophony toward alkali; shellac when dissolved in alcohol is capable of neutralizing a much smaller amount of caustic soda or potash than rosin when under similar conditions, and tests based upon this property are used and are of some value particularly as corroborating the results of other tests. The difference, however, between shellac and rosin in this respect is not sufficiently marked nor are the various grades of pure shellac or of rosin sufficiently constant in their behavior when tested by such methods to furnish a fairly satisfactory method of analysis. (See Allen, Commercial Organic Analysis, Vol. II, Part 3, pp. 190–195.)

Another property of rosin which has been used to distinguish and determine it in admixture with shellac is the solubility in ether of its compound or salt of rosin with silver while the similar compound of silver with shellac acids remain undissolved. The neutral constituents of shellac, that is, those which generally unite with either soda or silver are, however, also soluble in ether and are likely to be obtained in admixture with the rosin so that this method of analysis is objectionable.

Shellac and rosin when treated with proper solutions of iodine behave very differently; the shellac absorbs a relatively small amount, varying from 7 to 18 per cent. of its weight of iodine, the amount depending upon the details of the process used, while rosin under similar conditions absorbs from 120 to 230 per cent. of its weight of iodine. For many years it was the practice to use for the examination of shellac in this way the Hubl process in which the iodine and the rosin to be tested are used in solution in alcohol to which a little chloroform has been added. This Hubl process was originally designed for the examination of fats and oils and although it was in its day a very useful process it has in recent years been replaced by others which are more rapid and accurate.

The process which has been for the past few years in most common use in the United States for the determination of rosin in shellac is the iodine process which was originally suggested for use upon fats and oils by Wijs and which was later studied and adapted for use upon shellac by Langmuir and was finally recommended by the Sub-committee on Shellac Analysis of the American Chemical Society in Journal of American Chemical Society, XXIX, 1221 to 1227, as the most reliable of those in use up to that time. The process consists in brief of the following steps: in a treatment of a fixed amount, 200 milligrams of the shellac to be tested dissolved in 20 c.c. of acetic acid of 99 per cent. strength to which 10 c.c. of chloroform is added, with 20 c.c. of a solution of iodine monochloride in acetic acid of 99 per cent. strength for exactly one hour at a temperature of 21 to 24° C. The amount of iodine which is absorbed under these conditions by rosin is assumed to be 228 per cent. of the weight of the rosin while shellac absorbs less than 18 per cent. of its weight and in making the calculation the iodine figure of the shellac is assumed to be 18. This process has given excellent results in practice particularly as it is capable of giving in the hands of different operators closely agreeing results upon the same sample. The objections to the use of this process are first that it is likely to give results below the truth and second that the method of ascertaining the amounts of rosin is an indirect one and any other substance having a high iodine figure would be counted as rosin; the rosin itself is at no stage of the process separated from the shellac and submitted to a separate examination.

The need of a process of analysis which would actually separate the rosin from the shellac so that it can be examined by itself has led the writer to devise a process which has recently begun to be used in which the rosin is so separated from the shellac. This process which was described at length in the "Journal American Chemical Society," XXX, 867 to 872, depends upon the fact that rosin is soluble in petrolic ether while shellac is not. Although it is not practicable to extract the rosin from solid shellac with petrolic ether, the latter may be used to separate the two resins when they are dissolved in a suitable solvent. The shellac to be examined is in this process dissolved in absolute alcohol or in glacial acetic acid; with both of these solvents petrolic ether is miscible. It is, therefore, added to the solution of the rosin, and to the resulting mixture water is added. This results in a separation of the alcoholic and petrolic ether layers as the diluted alcohol is no longer miscible with the petrolic ether. The petrolic ether carries with it the rosin, the wax contained in the shellac, providing sufficient petrolic ether was used to dissolve it, and traces of shellac or of some constituent of shellac. Several methods of separating the shellac wax from the rosin may be used, but the most convenient is to extract the petrolic ether solution of the two with an alkaline solution which removes the rosin and leaves the wax dissolved in the petrolic ether. From the alkaline solution of the rosin the latter may be recovered by acidifying, extracting the acidified solution with a solvent such as ether and distilling off the ether to obtain the rosin which may then be weighed.

After considerable experience with this method the following details as to mode of procedure have been found convenient as well as tending to give accurate results. Place 2 grams of the shellac to be examined in a 16-oz. flask and add to it 20 c.c. of absolute alcohol; dissolve the shellac in the alcohol by gentle heating. Now add slowly and with constant agitation 100 c.c. petrolic ether boiling below 80° C. The first addition of petrolic ether to the alcoholic solution does not occasion the precipitation of any shellac, but as further quantities are added the mixed solvent of alcohol and petrolic ether becomes incapable of retaining the whole of the shellac in solution and it gradually precipitates out. It is necessary that the addition of the petrolic ether should therefore be made slowly and with stirring in order that the precipitating shellac may not carry out with it mechanically any of the rosin contained in the solution. When the whole of the 100 c.c. of petrolic ether has been added 100 c.c. of water is added also with agitation. The first additions of water cause the separation of the liquid into two layers one of which is rather strong alcohol, which may dissolve some part of the rosin which is afterward precipitated by the rest of the water. The necessity for agitation during this stage is to insure the collection into the petrolic ether of all the rosin. When all the water has been added the liquid is poured into a tapped separator and the flask rinsed out with a little more petrolic ether. The two liquids in the separator readily separate and the watery layer is drawn off. The petrolic ether is then washed with a little water which is also drawn off. The petrolic ether is then filtered into a clean separator and to it is added 25 c.c. of a solution of N/5 caustic soda in 50 per cent. alcohol. This caustic soda solution is measured into the separator accurately with a pipette and a similar pipette full of the same solution is titrated with standard hydrochloric acid, using methyl orange as an indicator. The separator containing the alcoholic soda solution and the petrolic ether is then thoroughly agitated, allowed to settle, and the watery layer drawn off into a tared flat, bottomed dish. The petrolic ether is then washed by agitating with a little 50 per cent. alcohol and the washings are added to the tared dish. There is further added to the contents of the tared dish the same volume of the same standard hydrochloric acid as were required by the check portion of 25 c.c. of the N/5 soda. In this way the entire quantity of soda is neutralized with hydrochloric acid and the resinous matters contained in the solution are left uncombined. The contents of the dish are then allowed to evaporate at a low temperature until the residue is dry, when the dish with its contents is weighed. The check portion of 25 c.c. of N/5 soda is, after being neutralized with standard acid as described, evaporated in a similar dish and in this way the amount of sodium chloride produced from the 25 c.c. portion is ascertained. By deducting the weight of sodium chloride so found from the combined weight of the resinous matter and sodium chloride the weight of the former is ascertained. The dried contents of the dish may now be further examined. The odor of the resin and its consistency may be observed, its acidity may be determined by solution in neutral alcohol and titration with N/10 alkali or if it is desired the resinous matter may be separated from the sodium chloride by dissolving it in ether or some other solvent.

By this process it is practicable to actually separate, examine, and exhibit the rosin which had been added to the shellac as an adulterant. It is also practicable to determine the amount of wax present providing, however, that a much larger amount of petrolic ether had been used than is necessary for the complete extraction of rosin alone. It has been found necessary to use at least 200 c.c. of petrolic ether per gram of shellac in order to insure the complete extraction of the wax, while 50 c.c. of petrolic ether per gram of shellac appears quite sufficient to extract any reasonable amount of rosin.

As previously stated the petrolic ether dissolves traces of pure shellac. It becomes important, therefore, to know how much will be dissolved from pure shellac when examined by the process above described. A great many pure shellacs have therefore been examined by this process and at the same time their iodine figures have been ascertained by the Wijs-Langmuir process. The following figures are representative of the results obtained:

Grade	Iodine figure	% Extracted matter	Acidity of extracted matter per 2 grams in c.c. N/10 KOH.
Angelo-B-Pure Button.	14.7	2.12	1.6
Pure T-N	18.0	1.79	1.5
Pure T-N	17.7	1.05	0.8
Pure T-N	16.0	2.19	1.6
Pure T-N	15.9	2.03	1.5
Pure T-N	17.6	1.82	1.5
Pure orange	17.4	1.38	1.2
Sticklac..............	14.7	3.07	2.3

It will be noted that there seems to be a tendency for the higher grades to give a larger amount of rosin soluble in petrolic ether than the lower grades. The amount given by such grades of shellac as are likely to be found adulterated may safely be assumed to be less than 2 per cent. If the shellac is of the highest grades, those which seldom contain rosin and which consequently seldom need to be examined by the analyst, the amount of soluble matter may with safety be assumed to be less than 3 per cent.

The rosin which is to be determined consists principally of acid bodies which readily unite with caustic soda, but it also contains a certain amount of unsaponifiable matter which is not extracted from the petrolic ether by the soda solution. The amount of rosin which is finally weighed cannot, therefore, be greater than the amount of free acids originally contained in it. The amount of unsaponifiable matter contained in the rosin used is, therefore a matter of interest, but unfortunately it is impossible to ascertain the exact amount. It is reasonable to assume, however, that not more than 85 per cent. of the rosin will be recovered and weighed in the process above described. This assumption agrees perfectly with the experience in this laboratory in working upon a considerable variety of rosins.

In calculating the amount of rosin content from the weight of extracted resinous matter in this process, allowance must be made for the small amount of pure shellac extracted and also for the shortage in the weight of rosin as finally weighed on account of the unsaponifiable matters contained in it.

In making this calculation the following formula is used:

If Y = per cent. rosin,
M = per cent. of extract of pure shellac,
N = per cent. of extract of pure rosin,
A = per cent. of extract of mixture,

Then $Y = \frac{100 (A-M)}{N-M}$

Here M = 2.0 or in the case of high grade shellac, 3.0, and N = 85.0.

Then $Y = \frac{100 (A-2)}{83}$.

Shellac varnishes may contain beside true shellac not only rosin, but other gums and resins soluble in alcohol. It becomes, therefore, a matter of interest to ascertain how some of these other resins behave when treated by this process. Two samples of manilla, when treated, using absolute alcohol as the first solvent, gave, respectively, 41.2 and 43.3 per cent. of matter soluble in petroleum ether. The acidity of these two lots of matter soluble in petroleum ether was in the case of the first sample such that 1 c.c. of normal alkali neutralized 411.7 mg. and in the case of the second 470.7 mg. Two samples of Kauri gave, respectively, 37.9 and 27.0 per cent. Upon titrating with standard alkali these portions soluble in petroleum ether, it appeared that 1 c.c. of normal alkali was capable of neutralizing 903.6 mg. and 742.5 mg., respectively. Of Sandarac, two samples, when similarly analyzed, gave 34.96 and 36.19 per cent., having such an acidity that of the first 541.2 mg. would neutralize 1 c.c. normal alkali, and of the second 552.5 mg. would neutralize 1 c.c. Of Dammar, 89.9 per cent. proved to be soluble, while the resin of *Shorea roburta*, a sample of which was kindly sent by Mr. W. Risdon Criper, of Calcutta, gave 69.5 per cent. of soluble matter.

APPENDIX A.

THE ANALYSIS OF BITUMINOUS PAINTS.

At the present time many bitumens and artificial bitumens are frequently used, either alone or in combination, in the manufacture of paints, black varnishes, and japans. The asphaltic compounds are naturally occurring products in many cases containing comparatively large percentages of sulphur. Mineral matter, which is present in widely varying roportions, consists usually of limestone, clay, or sandstone, containing the usual impurities found in these materials.

Petroleum residuum and coal-tar pitch are sometimes used alone as paints, but more frequently petroleum residuum is added to asphaltic compounds as a flux. Free carbon also finds application as a color agent for deepening such mixtures, but experience has shown that the percentage of free carbon should not exceed 20 per cent.

While a chemical analysis of such mixtures will disclose little concerning their true value as paints, nevertheless it is in many cases advisable and necessary that an examination be made so as to determine their general composition. The following methods will be found to give good approximate results in the examination of paints made from asphaltic and bituminous compounds.

Separation and Determination of Volatile Constituents.—100 grams of the paint, after being thoroughly mixed, are placed in a distilling flask and the volatile constituents separated in the usual way, as described on page 66, all the volatile constituents distilling over up to 180° C. being carefully collected. The amount of distillate thus found will give the percentage of volatile constituents present. The distillate is then fractionated so as to determine the percentage of benzol, benzene, turpentine, and volatile oils. This operation is carried out by the usual fractionation methods, fractions being weighed and the percentage of each constituent determined. Separation and estimation of the turpentine in the distillate is best done by the polymerization method described on page 62.

Any water present in the vehicle or in the asphalt itself, as so frequently occurs in asphaltic mixtures, will distil over at the above temperature and will be found in the distillate.

It must also be understood that some low-boiling oils which are so often present in bituminous mixtures may distil over at or below 180° C. When such oils are present, it is impossible to differentiate between them and the oils present in the vehicle. In such cases, however, they may be assumed to act in the rôle of vehicle and may be reported as such.

Nonvolatile Residue.—The residue left in the flask after distillation is then examined for bituminous matter (petrolene and asphaltene), nonbituminous matter, carbon, sulphur, and the constituents entering into the ash.

Petrolene and asphaltene are the names applied to the substances obtained by the use of certain solvents on the natural and artificial bitumens. The portion soluble in petroleum spirit, ether, or acetone is known as petrolene, while the portion insoluble in petroleum spirit but dissolved by boiling turpentine and cold chloroform is known as asphaltene.

The total bituminous matter is considered to be that portion of the asphaltic mixture which is soluble in the above solvents or in carbon bisulphide, the portion insoluble being considered as nonbituminous matter. Pure asphalt, as used in the manufacture of painting materials, should be completely soluble (with the exception of 4 or 5 per cent. which may be mineral matter) in carbon disulphide, oil of turpentine, or chloroform.

Nonbituminous Matter, Carbon, and Ash.—The residue remaining after the extraction by the carbon bisulphide method is dried and weighed. The loss between the original weight of the nonvolatile residue and the weight of the unextracted matter, gives the percentage of bituminous matter (asphaltene and petrolene) present. The insoluble residue after weighing is ignited until all the carbon has been burned off. The weight is again taken and the loss reported as nonbituminous matter and carbon. It is advisable before weighing to treat the ash with a little ammonium carbonate to reconvert any carbonates which may have been decomposed by the heating to their original form.

Ash.—The constituents of the ash are determined in the

manner outlined under the analysis of mixed pigments, or by any of the methods used in the analysis of limestones, clays, or sandstones.

The mineral matter present quite frequently exists in combination with the bituminous matter forming resinous-like compounds which partially dissolve in the solvents used for fractionation of the nonvolatile residue, thus giving high results for the bituminous matter and low results for ash. These errors, however, serve to counterbalance each other and are frequently so inappreciable that they may be neglected.

The correct percentage of ash should be determined by igniting a new sample of the nonvolatile residue until all the carbon has been destroyed.

Sulphur.—It is often necessary that the sulphur content of asphaltic mixtures be determined. This may be approximately determined by following Eschka's method:

Mix intimately 1 gram of the finely ground, nonvolatile residue with 1 gram of calcined magnesium oxide and .5 gram of mixed sodium-potassium carbonates, in a platinum or porcelain crucible. After thorough mixing, the crucible (uncovered) is heated to a dull red heat with an alcohol or Bunsen flame, in the latter case the crucible being placed in a hole cut into an asbestos board, thus preventing any sulphur from the flame from contaminating the mixture. The action is hastened by frequent stirring of the mixture. The heating is continued until the contents of the crucible become a dull yellow. Cool the crucible and mix the contents intimately with about 1 gram finely powdered ammonium nitrate. Heat until the ammonium nitrate is completely decomposed. Any sulphites formed by the first treatment are thus completely converted into sulphates.

The contents of the crucible, after cooling, are carefully transferred to a beaker and extracted with hot water. Evaporate, wash, acidify with hydrochloric acid and precipitate the sulphate present in the usual way with barium chloride. Weigh as barium sulphate and calculate to free sulphur. It has been found that the sodium peroxide method is apt to give results which are low.

The Hempel-Graefe * method for determining the percent-

* J. of Ind. and Eng. Chem., May, 1910, p. 187.

age of sulphur in bitumens or pyro-bitumens, consists in burning a small quantity of the material under examination, in an atmosphere of oxygen, with absorption of the gas in sodium peroxide. Subsequent neutralization and precipitation is made.

In determining the presence and identification of vegetable or fossil gums, such as rosin or kauri gums, the methods of McIlhiney for the analysis of shellac together with the methods given for the analysis of varnish will prove useful.

Examination of the Nonvolatile Residue.

A portion of about 50 grams of the nonvolatile residue is placed in an Erlenmeyer flask and shaken with a considerable quantity of carbon bisulphide. After sufficient time has elapsed for the carbon bisulphide to exert its solvent power on the petrolene and asphaltene, the contents of the flask are poured upon a suitable filter. If pigments are present, they will be found upon the filter and may be examined by the methods outlined for the analysis of pigments in mixed paints.

If drying oils, such as linseed oil, are present in the nonvolatile residue, they may be removed by treating another portion of the residue with 88° gasolene for eight or ten hours. Evaporation of the filtrate from this product will leave the oils which may be present in a condition suitable for analysis. Rosin and resinates are also extracted by this treatment and should be looked for. Fossil gums, however, are apt to be precipitated by the gasolene.

FORREST METHODS.

The Examination of Black Varnishes and Enamels.

The following method for the examination of paints containing bitumens or pyro bitumens has been prepared for this book by Mr. C. N. Forrest, Chief Chemist of the New York Testing Laboratory, where extensive tests on the nature of bituminous materials have been conducted. These methods are of great importance and serve to give much new information regarding Bituminous Paints.

For the purpose of an analysis, bituminous paints should be divided into three classes, as follows:

Asphaltum varnish.
Asphaltum enamel.
Coal-tar enamel.

Asphaltum varnish, as its name implies, does not contain a pigment but consists of a base of asphatum and other substances, reduced to a fluid consistency with a suitable volatile solvent.

Asphaltum enamel consists of an asphaltum varnish in combination with a pigment, and may be either black or colored.

Coal-tar enamel, although generally considered as a varnish or paint, always contains carbon in suspension and therefore should be classified as an enamel.

A distinctive feature of bituminous varnishes and enamels is that they dry by evaporation rather than by oxidation. If a drying oil is present a certain degree of hardening of the film subsequently occurs, but the intinal drying of such varnishes and enamels depends upon the spontaneous evaparation of the volatile thinners present, and the rapidity of drying upon the degree of volatility of the volatile solvent. Any linseed or other oil present should be considered as a constituent of the base.

An asphaltum varnish will therefore consist of a basic material dissolved in from 25 to 60 per cent. of some volatile solvent, such as turpentine, benzine, heavy petroleum spirit, or coal-tar spirit. Occasionally carbon disulphide or some special solvent may be employed, but that would be unusual.

An asphaltum enamel will contain black or colored pigments combined with a varnish of essentially the same nature as has just been described.

A coal-tar enamel will consist essentially of a base of coal-tar pitch dissolved in benzole or other coal-tar spirit. The free carbon present is very finely divided and will remain suspended for an indefinite period.

There are several kinds of hard asphaltum available for varnish-making, but the principal and most generally used types are mentioned in the following table, which also gives the important characteristics of the same.

	Gilsonite.	Nanjak.	Grahamite.	Refined Bermudez asphalt.	Oil pitch.
Specific gravity at 77° F	1.049	1.0844	1.171	1.0575	1.0703
Color of powder....	Brown.	Dark brown.	Black.	Black.	Black.
Melting-point......	325° F.	350° F.	Intumesces.	170° F.	164° F.
Bitumen soluble in CS_2	99.9%	99.2%	94.1%	96.0%	98.2%
Mineral matter	.1	.3	5.7	2.0	Trace.
Difference	.0	.5	.2	2.0	1.8
	100.0	100.0	100.0	100.0	100.0
Bitumen soluble in 88° naphtha	15.9	26.9	.4	69.1	69.6
This is per cent. of total bitumen....	15.9	27.0	.4	71.9	70.9
Residual coke.....	13.4	25.0	53.3	14.0	19.5
Paraffine	None.	None.	None.	None.	.8
Characteristics of asphaltenes insoluble in 88° naphtha.					
Color	Black.	Black.	Black.	Brown.	Brown.
Condition..........	Hard.	Hard.	Hard.	Soft.	Soft.
Residual coke	15-20%	55-60%	50-55%	30-35%	50-60%

The characteristics of the most select grades of coal-tar pitch suitable for the manufacture of enamel are as follows:

Coal-tar Pitch.

	Hard.	Soft.
Specific gravity at 77° F.	1.24	1.23
Color of powder	Black.	Black.
Melting-point	161° F.	140° F.
Bitumen soluble in CS_2	94.1%	92.2%
Mineral matter	.1	.1
Free carbon, etc.	5.8	7.7
	100.0	100.0
Bitumen soluble in 88° naphtha	85.0	87.0
This is per cent. of total bitumen	90.3	94.4
Residual coke	29.1	25.1

The characteristics of the most select grade of stearine pitch, Calabrea pitch, and of the resins sometimes included in black varnishes are as follows:

CALABRIA PITCH.

Specific gravity at 77° F.	1.030
Color of powder	Black.
Melting-point	161° F.
Bitumen soluble in CS_2	99.5%
Mineral matter	.5
	100.0
Bitumen soluble in 88° naphtha	41.3
This is per cent. of total bitumen	41.5
Residual coke	11.5
Characteristics of asphaltenes insoluble in 88° naphtha.	
Color	Black.
Condition	Spongy—melts.
Residual coke	21.8%

Resins.

	Colophony.	Kauri.	Copal.
Soluble in 88° naphtha	100%	4.0%	25.0%
Residual coke	Trace.	Trace.	Trace.

There is no established custom followed in the selection of the other materials which are combined with hard asphaltum in the preparation of the base of a varnish.

Such materials include rosin, fossil gums, linseed oil, China wood-oil, mineral oil, driers, and other substances. The purpose of the use of such materials is to modify the degree of hardness and flexibility, facilitate the thinning operation and increase durability according to the service the finished varnish is to perform.

A coal-tar enamel is usually nothing more than a coal-tar pitch dissolved in coal-tar spirit. It may sometimes contain rosin or a small amount of linseed oil, but that is unusual. It is not customary to mix coal-tar pitch and asphaltum for a varnish base as the two materials do not combine readily and, if combined, are furthermore generally but partially soluble in petroleum spirit, the least costly solvent employed in this industry.

A practical test should always be made by applying a coat of the varnish or enamel on a surface similar to that upon which it is to be used and the covering or hiding capacity, spreading qualities, time of drying, etc., noted. An outdoor exposure test on a small steel plate of the materials intended for such use should also be made.

The first step in the analysis of bituminous varnishes and enamels should be to separate the volatile thinners from the base, and this is most conveniently done by distilling about 50 grams in a flask by gentle heat. A nonoxidizing atmosphere should be maintained in the flask during the distillation by the introduction of CO, or other inert gas. The difference in the boiling-point of the solvent and the base is so great that a clean separation may be readily made by this method.

The temperature at which the volatile thinners pass over should be noted and a subsequent examination of the distillate for volume, specific gravity, flash-point, and indifference to strong sulphuric acid will give sufficient data for the identification of the same and the amount present.

While the base in the distilling flask is still fluid it should be poured into a shallow tin box. The portion adhering to the flask may be removed with carbon disulphide, the solvent expelled, and the residue combined with the chief portion. This precaution, however, need not be observed unless a pigment is present.

The amount of pigment may be determined by treating 10 to 50 grams of the base with 100 c.c. carbon disulphide in a small Erlenmeyer flask and filtering through a Gooch crucible, washing with the same sovlent, or by extracting with carbon disulphide in a Soxhlet.

Having separated the pigment as above and recovered the soluble portion by evaporating off the solvent, the analysis should proceed exactly as in the case of a clear asphaltum varnish base.

From 1 to 5 grams of the base, in a finely divided condition, should be treated in an Erlenmeyer flask with 100 to 200 c.c. 88° naphtha and allowed to stand over night at laboratory temperature. The naphtha is then decanted through a Gooch or filter-paper, the residue transferred to the filter and washed with the solvent until the filtrate runs through clear.

The filtrate will contain the rosin drying and petroleum oils and a portion of the asphaltum or pitch if any is present.

The insoluble residue will consist essentially of the so-called asphaltenes of the asphaltum, and will generally represent from 30 to 85 per cent. of the amount of asphaltum in the base, depending upon what variety of the same has been employed. In the case of Grahamite, it would represent practically all of the asphaltum present, but that material is not in general use in varnishes.

The color and hardness of the insoluble residue should be noted and a determination of residual coke or fixed carbon made as follows:

One gram of residue is ignited in a platinum crucible, as in the proximate analysis of coal, Jour. Amer. Chem. Soc., Vol. 21, p. 1116, 1899.

If the residue is dark brown and soft and has a fixed carbon of about 50 per cent. the asphaltum is probably an oil pitch.

If it is black and hard and has a fixed carbon of about 50 per cent. it is probably grahamite. If black and hard and about 15 per cent. fixed carbon it is gilsonite. If dark brown and soft and about 30 per cent. fixed carbon it is probably Bermudez asphalt.

The filtrate after expelling the solvent used in the extraction may be tested for saponification and iodine values and for rosin by the Lieberman or Storch test. If a quantitative separation of rosin is desired the laborious Twitchell method must be resorted to.

The difference between the saponifiable and the total soluble in 88° naphtha will give the mineral oils and so-called petrolenes of the asphaltum.

A positive test for and approximate determination of coal-tar may be made by distilling a small amount of the base in a glass retort to coke, and mixing 4 c.c. of the distillate thus obtained with 6 c.c. dimethyl sulphate in a graduated tube. Coal-tar distillates are entirely soluble in dimethyl sulphate, while those from asphaltum, petroleum, and vegetable paint oils and resins are insoluble. See Jour. Ind. and Eng. Chem., Vol. II, p. 186, May, 1910.

The pigment separated from asphaltum enamels may be examined in detail if desired by the same methods employed in the analysis of linseed oil and house paint.

The combinations in use by various manufacturers in the making of black varnishes are frequently very complicated.

Baking enamels for certain specific uses require speicia formulæ, but for the general line of black varnishes the princpall object is to produce a base of high gloss and deep color, having some flexibility, which will reduce with turpentine or benzine at a temperature sufficiently low to avoid excessive loss of solvent in the manufacturing process.

To this end it is usually sufficient to fuse an asphaltum of the gilsonite type with rosin, linseed, or mineral oil in sufficient amount to reduce the melting-point of the asphaltum without rendering it unduly soft.

The durability of the film of varnish is affected by the fluxing material used as well as its degree of hardness and flexibility, and the desire to produce a cheap varnish frequently overbalances the desire to produce a durable one.

The base of a varnish for outdoor exposure to sunlight and atmospheric conditions should be of a different character both in degree of hardness and composition from varnishes intended for damp-proofing, insulating, and acid, and alkali resisting purposes.

There is nothing to be gained by the use of turpentine and other expensive solvents instead of benzine or heavy petroleum spirit, provided the base is entirely soluble in the latter and a proper selection has been made to assure the drying properties desired. Varnish makers' benzine (62°B.) evaporates too fast to be satisfactory as the thinning material of a heavy brush varnish. On the other hand, it is indispensable in thin quick-drying brush varnishes and dips.

On account of the extremely complex nature of black varnish bases it is quite impossible to prescribe a general method which may be followed without modification and cover all features of this subject.

The foregoing is, therefore, given more as an outline for the guidance of those analysts who are capable of adapting the general schemes of classification and separation to the particular purpose at hand, and supplementing the same with such further tests as will develop the special features of the material if the simple separations, etc., specifically mentioned are not sufficient.

APPENDIX B.

The following specifications of the Army and Navy Departments are given so that the chemist engaged in the examination of such materials may become better acquainted with the requirements. These specifications, however, refer only to the chemical requirements, and for full specifications the reader is referred to the Bureau of Supplies and Accounts of the United States Navy, Washington, D. C.

SPECIFICATIONS ISSUED BY THE ARMY DEPARTMENT.

Pure White Lead.—White lead must be of the best quality, finely ground in pure well-settled raw linseed oil; must be of maximum whiteness; must work freely under the brush, and not be crystalline in structure nor deficient in density and opacity. Dry pigment must contain at least 98 per cent. of hydrate carbonate of lead. Its workings under the brush, maximum whiteness, body, and covering qualities to be determined by practical test.

White **Zinc.**—*American Process.*—The dry pigment must contain at least 98 per cent. of oxide of zinc, not more than 0.5 per cent. of sulphur in any form, and be of the quality known as "XX."

French Process.—The dry pigment must contain at least 99 per cent. of oxide of zinc and not more than 0.25 per cent. of sulphur in any form, and to be of maximum whiteness as compared with standard sample.

Venetian Red.—The dry pigment must contain at least 40 per cent. of sesquioxide of iron, not more than 15 per cent. of silica, the balance to consist of sulphate of lime that has been fully dehydrated by dead burning and rendered incapable of taking up water of crystallization.

Indian Red.—Must be of good rich color. The dry pigment must contain at least 95 per cent. of oxide of iron (Fe_2O_3), and be free from sulphur and alkali. Pale shade is desired in the absence of other specifications.

Vermilion.—*American* (*dry*).—Must be of good, bright color, and contain at least 98 per cent. of basic chromate of lead, and be free from any foreign coloring matters.

English (*dry*).—Must contain at least 99 per cent. of red sulphide of mercury; must be free from any foreign coloring matters or alkali in any form.

Artificial.—The dry pigment must be the lead-barium lake of the azo dye known commercially as "Lithol."

Raw and Burnt Sienna.—The dry pigment must be equal in quality to the best selected Italian sienna, and must not contain more than 5 per cent. of lime in any form.

Yellow Chromes.—Must be of good bright color and full strength. The dry pigment must contain at least 98 per cent. of normal chromate or basic chromate of lead.

Yellow Ochre.—It must be equal in color and quality to the best French ochre and be free from any chromate of lead or foreign coloring matter. The dry pigment must contain at least 20 per cent. of oxide of iron and not more than 5 per cent. of lime in any form.

Chrome Green.—Chrome green must be of good bright color. The dry pigment must contain 25 per cent. chrome green made by mixture of pure chrome yellow and Russian blue and 75 per cent. barium sulphate. Medium shade is desired in absence of other specifications.

Drop Black.—Drop black must be of good deep luster and consist of calcined bone-black only. The addition of blue or gas carbon-black will be ground for rejection. The paste must contain not less than 45 per cent. of pure pigment.

Oil Lamp-black for Tinting Purposes.—The pigment must be the perfectly calcined product of oils only and show less than 2 per cent. of ash. It must be absolutely neutral, free from oil or greasy matter, grit, and all impurities. The pigment reduced in white must give a clear blue-gray tone or tint.

Japan Drier.—Japan drier must not flash below 103° F. (open tester); must be of the best quality and made from pure kauri gum, pure linseed oil, pure turpentine and the proper driers only; must set to touch in from one-fourth to one hour, dry elastic in from eighteen to twenty-four hours at a temperature of 70° F., and must not rub up or powder under friction by the finger. When mixed with pure raw linseed oil in the proportion of eight parts of oil to one part of drier must remain clear for two hours and set to touch in from six to seven hours at a temperature of 70° F.

Varnish.—All varnishes other than those which have definite specifications must be pure turpentine hard-gum varnishes and absolutely free from rosin or any turpentine substitutes.

Damar Varnish.—It must be made from solution of the very best quality of damar gum; such solution to contain at least 50 per cent. of gum with 45 per cent. turpentine. It must be digested cold and well settled. It must be as clear as and not darker than the standard sample. It must be free from benzine, rosin, and lime or other mineral matter. Its specific gravity at 60° F. must be between .935 and .937 and its flash point between 105° and 115° F. It must set to touch in not more than twenty minutes, and when mixed with pure zinc oxide must show a smooth glossy surface equal to that shown by the standard sample.

Tests.—Besides chemical tests to determine the above qualities and practical tests to determine its qualities of finish, a board properly coated with a mixture of zinc and the liquid will be exposed to the weather for a period of one month, and at the end of this time must have stood such exposure equally as well as the standard sample. A similarly prepared sample will also be baked at 250° F., and must not at this temperature show any greater signs of cracking, blistering, or any other defects than standard samples under the same conditions.

Asphaltum Varnish.—Asphaltum varnish must be made of pure high-grade asphaltum of the very best quality, of pure linseed oil and pure turpentine dryers only, and must not contain less than 20 gallons of prepared linseed oil to 100 gallons of varnish. It must not flash below 103° F. (open tester). It must mix freely with raw linseed oil in all proportions; must be clear and free from sediment, resin, and naphtha. When flowed on glass and allowed to drain in a vertical position the film must be perfectly smooth and of full body, and must equal in this last respect the standard sample. It must set to touch in from one and one-half to two and one-half hours and must dry hard in less than twenty hours at 70° F. When dry and hard it must not rub up or powder under friction by the finger. The application of heat must quicken the time of drying and give a harder film.

SPECIFICATIONS ISSUED BY THE NAVY DEPARTMENT.

White Lead in Oil.—The dry pigment must be of the best quality, must not be crystalline in structure or deficient in density or opacity. Unless otherwise specified, white lead will be delivered in paste form, the pigment finely ground in pure raw linseed oil, and in paste form must not contain more than 0.5 per cent. of moisture. The dry pigment must be a pure hydrated carbonate of lead, free from all adulterants, and equal in quality to the best commercial grades. The total acetate must not be in excess of the equivalent of 0.15 per cent. of absolute acetic acid.

Specifications for Whiting.—The material must be free from grit; must contain not more than 1.5 per cent. of matter insoluble in dilute hydrochloric acid, and not more than 4/10 per cent. oxides of iron and aluminum (determined together).

Specifications for Chrome Green.—The dry pigment must be of a good bright color and must contain at least 98 per cent. by weight of pure lemon chrome and Chinese blue, which mixture must not contain more than 10 per cent. by weight of lead sulphate and must be equal in all respects to the standard sample. A medium shade is desired in the absence of other specifications.

Metallic Brown.—1. The dry pigment must contain not less than 45 per cent., by weight, of oxide of iron and must not contain more sulphur in combination than the equivalent of 2 per cent., by weight, of sulphur trioxide (SO_3).

2. It must be ground perfectly pure, not made from or adulterated with the by-products of sulphuric acid works, and must be free from makeweights or adulterants. When in paste form, it must contain at least 20 per cent., by weight, of pure raw linseed oil.

Chinese Blue.—1. The dry pigment must contain not less than 98 per cent., by weight, pure coloring matter of the best quality, free from adulterants, and equal in every respect to the standard sample.

2. When in paste form, the paste must contain not less than 50 per cent. by weight of pure pigment ground in absolutely pure, well-settled, and perfectly clear raw linseed oil of the best quality only to a medium stiff consistency, which will break up readily in thinning, and must be free from grit, adulterants, and all impurities.

Red Lead, Dry.—The dry pigment must be of the best quality, free from all adulterants, and contain at least 94 per cent. of true red lead(Pb_3O_4)—equivalent to 32.8 per cent. of lead peroxide (PbO_2)—, the balance to be practically pure lead monoxide (PbO). It must contain less than 0.1 per cent. of metallic lead, and to be of such fineness that not more than 0.5 per cent. remains after washing with water through a No. 21 silk bolting cloth sieve. It must be of good bright color and be equal to the standard sample in freedom from vitrified particles and in other respects.

Chrome Yellow (Medium Orange).—1. The dry pigment must be of good bright color and full strength, and must contain at least 98 per cent. by weight of normal chromate or basic chromate of lead.

2. The pigment must be of the best quality, finely ground in absolutely pure, well-settled, and perfectly clear raw linseed oil of the best quality only to a medium stiff paste, which will break up readily in thinning, and must be free from grit, adulterants, and all impurities.

Specifications for Spar Varnish.—1. To be of the best quality and manufacture and equal in all respects, including body, covering properties, gloss, finish, and durability, to the standard sample in the general storekeepers' offices at the various navy yards. To be made exclusively from the best grade of hard-varnish resins, pure linseed oil, pure turpentine, and lead-manganese driers, and to be free from all adulterants or other foreign materials.

2. The varnish must not flash below 105° F. (open tester), and when flowed on glass must set to touch in from six to twelve hours and dry hard in from thirty to forty-eight hours at a temperature of 70° F. To be as clear as and not darker than the standard sample, and to be equal to it in all respects as above specified.

Interior Varnish.—1. To be of the best quality and manufacture and equal in all respects, including body, covering properties, gloss, finish, and durability, to the standard sample in the general storekeepers' offices at the various navy yards; to be made exclusively from the best grade of hard-varnish resins, pure linseed oil, pure spirits of turpentine, and lead-manganese driers, and to be free from all adulterants or other foreign materials.

2. The varnish must not flash below 105° F. (open tester), set to touch in from six to eight hours and dry hard within twenty-four hours in a temperature of 70° F. It must stand rubbing with pumice-stone and water in thirty-six hours without sweating, and must polish in seventy-two hours with rottenstone and water; to be as clear as and not darker than the standard sample and to be equal to it in all respects, as above specified.

Japan Drier.—1. Japan drier must not flash below 105° F. (open tester); must be of the best quality, light in color, and be made from pure kauri resin, pure linseed oil, pure spirits of turpentine, and lead-manganese driers, and be free from adulterants, foreign material, sediment, and suspended matter. When flowed on a glass plate and allowed to drain in a vertical position the material must not come off when touched lightly with the finger after from fifteen to sixty minutes, and must dry elastic in not less than eight hours nor more than twenty-four hours at a temperature of 70° F., and must not rub up or powder under friction by the finger at the end of this time. When mixed with pure raw linseed oil (that will not break under 600° F.) in the proportion of eight parts of oil to one part of drier the mixture must remain clear for at least two hours, and when flowed on a glass plate must not come off when touched lightly with the finger at the end of eight hours at a temperature of about 70° F.

Raw Linseed Oil.—Must be absolutely pure well-settled linseed oil of the best quality; must be perfectly clear and not show a loss of over 2 per cent. when heated to 212° F., or show any deposit of foots after being heated to that temperature. The specific gravity must be between 0.932 and 0.937 at 60° F.

To be purchased by the commercial gallon; to be inspected by weight, and the number of gallons to be determined at the rate of 7 1/2 pounds of oil to the gallon.

Boiled Linseed Oil.—Must be absolutely pure kettle-boiled oil of the best quality, and the film left after flowing the oil over glass and allowing it to drain in a vertical position must dry free from tackiness in twelve hours at a temperature of 70° F.

It must contain no resin. The specific gravity must be between 0.934 and 0.940 at 60° F.

To be purchased by the commercial gallon; to be inspected by weight, and the number of gallons to be determined at the rate of 7 1/2 pounds of oil to the gallon.

Spirits of Turpentine.—1. The turpentine must be the properly prepared distillate of the resinous exudation of the proper kinds of live pine or live pitch pine, unmixed with any other substance; it must be pure, sweet, clear, and white, and must have characteristic odor.

2. A single drop allowed to fall on white paper must completely evaporate at a temperature of 70° F. without leaving a stain.

3. The specific gravity must not be less than 0.862 or greater than 0.872 at a temperature of 60° F.

4. When subjected to distillation, not less than 95 per cent. of the liquid should pass over between the temperature of 308° F. and 330° F., and the residue should show nothing but the heavier ingredients of pure spirits of turpentine. If at the beginning of the operation it shows a distillation point lower than 305° F., this will constitute a cause for rejection.

5. A definite quantity of the turpentine is to be put in an open dish to evaporate, and the temperature of the dish will be maintained at 212° F.; if a residue greater than 2 per cent. of the quantity remains on the dish it will constitute a cause for rejection.

6. *Flash Tests.*—An open tester is to be filled within 1/4 inch of its rim with the turpentine, which may be drawn at will from any one can of the lot offered under the proposal. The tester thus filled will be floated on water contained in a metal receptacle. The temperature of the water will be gradually and steadily raised from its normal temperature of about 60° F. by applying a gas or spirit flame under the receptacle. The temperature of the water is to be increased at the uniform rate of 2° F. per minute. The taper should consist of a fine linen or cotton twine (which burns with a steady flame), unsaturated with any substance. When lighted it is to be used at every increase of 1° temperature, beginning at 100° F. It is to be drawn horizon-

tally over the surface of the turpentine and on a level with the rim of the tester. The temperature will be determined by placing a thermometer in the turpentine contained in the tester so that the bulb will be wholly immersed in the liquid. The turpentine must not flash below 105° F.

7. *Sulphuric Acid Test.*—Into a 30 cubic centimeter tube, graduated to tenths, put 6 cubic centimeters of the spirits of turpentine to be examined. Hold the tube under the spigot and then slowly fill it nearly to the top of the graduation with concentrated oil of vitriol. Allow the whole mass to become cool and then cork the tube and mix by shaking the tube well, cooling with water during the operation if necessary. Set the tube vertical and allow it to stand at the ordinary temperature of the room and not less than half an hour. The amount of clear layer above the mass shows whether the material passes test or not. If more than 6 per cent. of the material remains undissolved in the acid this will constitute a cause for rejection.

INDEX.

Spotted something that isn't quite right?

Let us know and we'll fix it for future reprints.
And as a thank you for your diligence, we'll
send you a free Sothis book of your choice.

publisher@sothis.press

This page intentionally left blank

Made in the USA
Monee, IL
05 November 2026

67808945R00069